# Science...
# *for Her!*

### MEGAN AMRAM

SCRIBNER

New York London Toronto Sydney New Delhi

SCRIBNER

An Imprint of Simon & Schuster, Inc.

1230 Avenue of the Americas

New York, NY 10020

First Scribner trade paperback edition November 2015

SCRIBNER and design are registered trademarks of The Gale Group, Inc.,
used under license by Simon & Schuster, Inc., the publisher of this work.

For information about special discounts for bulk purchases, please contact
Simon & Schuster Special Sales at 1-866-506-1949 or business@simonandschuster.com.

The Simon & Schuster Speakers Bureau can bring authors to your live event.
For more information or to book an event contact the Simon & Schuster Speakers Bureau
at 1-866-248-3049 or visit our website at www.simonspeakers.com.

Interior design by Brian Chojnowski

Photos on pages xxiv and 101 by Matthias Clamer for StocklandMartel.com

Photos on pages xxv, 1, 19, 36, 81, 84, 85, 121, 130, 143, 158, 164, and 189 © Liz Bretz

Photos on pages 19 and 51 © Alamy

Photo of white dwarf on page 71 © NASA, ESA, H. Bond (STScl), and M. Barstow
(University of Leicester)

Photo of neutron star on page 71 © NASA

Photos on pages 51, 68, and 127 © Gettyimages

Photo on page 128 © Shutterstock

Photo of Rachel Carson on page 172 from Library of Congress

Photos via Wikimedia Commons on pages 66, 71, 79, 112, and 173

Illustration on page 128 by Alexandra Rushfield

Manufactured in the United States of America

1   3   5   7   9   10   8   6   4   2

Library of Congress Control Number: 2014017574

ISBN 978-1-4767-5788-9
ISBN 978-1-4767-5789-6 (pbk)
ISBN 978-1-4767-5790-2 (ebook)

# Table of Contents

# Dedication (to all my besties!!!)

I dedicate this book to my best friends: Carly (my first best friend), Michelle (my best friend on long winter nights), Ali (my best friend when I'm super wasted), Nicki (my best friend when I'm sober, sometimes a girl's gotta rest so she can recharge and party more!), Ashley (my best friend when I'm with a famous person because she's way uglier than me so the famous person pays more attention to me), Katie (my best friend who loves Céline Dion so much it's crazy), Candy (my best friend when I have freckles), Kate (my best friend when my lupus flares up), Ellen (my best friend at 4:20 P.M., if you know what I mean ;)),  Miranda

FIG. 0.1

(my best friend who just had her *seventh* "miscarriage" but DO NOT TELL HER I TOLD YOU), Eliana (my best friend in synagogue), Mary (my best friend in Mass), Chloe (my best friend on every other weekend, I split custody with her other best friend Mel), Olive (my best friend when we're reading Sartre), Mel (my best friend who I get to gossip about Chloe with), 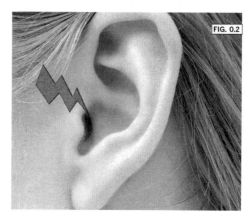 Amanda (my best friend when we're watching pro baseball), Jordan (my best friend who's so beautiful even though she's mixed race), Esther (my best friend who's my mom's best

FIG. 0.2

friend), Davida (my best friend who's Jewish), Mandy (my best friend when we're drinking La-Croix sparkling water, check it out online or in stores), Alexis (my best friend who like only wears rompers what's with that?), Carrie (my best friend who gets really jealous of how much I love Rachael Ray), Rachael Ray (I love you babe), Tiffany (my best friend who basically stole my boyfriend in the Lava Tube Caves), Heidi (my best friend when I want to feel fat, she is so skinny and perfect!), Minnie (my best friend who's literally a beautiful flower), Mary Katherine (my best friend in Catholic Mass when I'm closet-eating Communion wafers), Kristina (my best friend whose mom is dead RIP Nora), Nora (my best friend who's a dead mom),  Reverse Cowgirl (my best friend

FIG. 0.3

who is a sex move), Raita (my best friend who was my ex-boyfriend's tattoo teacher), *Marie Claire* (my best friend who is a magazine and a general composite of what the magazine industry sees as every modern woman), **FIG. 0.4** Clara (my best *frenemy*, she's a goddamn bitch, love ya, babe!), Adele (my best friend who may have committed vehicular manslaughter, innocent till proven guilty, bitch), Aisha (my best friend with the most best friends), Katja (my best friend from 1992–95, fuck that bitch), Lauren (my best friend

**FIG. 0.4**

whose titties look like mosquito bites), Claire-Marie (my best friend who has this amazing cuff bracelet, you should see it), Kathleen (my best friend who's Asian but like a cool Asian), Crystal (best friend to do meth with), Crystal Glass (other best friend to do meth with), Christina (other best friend to do meth with), Tina (other best friend to do meth with), Cris (other best friend to do

meth with), Cristy (other best friend to do meth with), Ice (other best friend to do meth with), Getgo (other best friend to do meth with), G (other best friend to do meth with), Trash (other best friend to do meth with), Super Ice (other best friend to do meth with), LA Glass (other best friend to do meth with), LA Ice (other best friend to do meth with), Ice Cream (other best friend to do meth with), Quartz (other best friend to do meth with), Chunky Love (other best friend to do meth with), Cookies (other best friend to do meth with), No Doze (other best friend to do meth with), Pookie (other best friend to do meth with), Rocket Fuel (other best friend to do meth with), Scooby Snax (other best friend to do meth with), Rebecca (my best friend who is a drug counselor who tried to help me quit meth—nice try but no cigar, Becs!), Marissa (my best friend who I may or may not have kissed in college), Carla (my best friend who I thought was my boyfriend but was actually a woman), **FIG. 0.5** Ella (my best friend when we're checking out the seniors at the local summer camp), Aubrey (my best friend whose blood type is B negative), Lily (my best friend who's a Queen Bitch judgette), Bridget (my best friend

**FIG. 0.5**

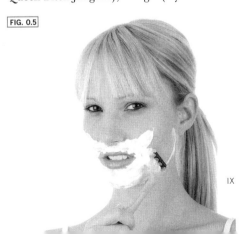

who has a beautiful soprano singing voice), Sofia (my best friend who has a beautiful mezzo-soprano singing voice), Grace (my best friend when we're buying capes online), Hannah (my best friend when we're buying capes in stores), Amelia (my best friend on my cheat day),  Arianna (my best friend who blogs), Harper (my best friend who's a sugar baby), Lillian (my best friend who would buy condoms for boys who didn't want to sleep with her), Charlotte T. (my best friend with a million siblings), Charlotte A. (my best friend with super curly hair), Charlotte M. (my best friend named Charlotte, sorry Charlottes T. and A.!), Evelyn (my best friend when I want to be head bitch for a change), Victoria (my best friend with a "secret" and the secret's chlamydia! Just

**FIG. 0.6**

playing with you, Vick, but for real she does have it), Brooklyn (my best friend when I'm in Manhattan), Zoe (my best friend when we're on a juice cleanse), Layla (my best friend when we're on a beef-stew cleanse),  Hailey (my best friend who looks so young because she still wears pig-

tails), Leah (my best friend who nailed which black blazer looks best on her body type), Kaylee (my best friend who believes that JFK was killed by a single shooter, what a fucking idiot), Riri (my best friend who gonna be in the Bee), Gabriella (my best friend who has no limbs and I honestly don't know how or why), Alison (my best friend

**FIG. 0.7**

who's short but wears really high heels so it totally tricks stupid people), Allison (my best friend who's a therapist with red hair), Shirlene (my best friend who has never dated anyone and never will because she isn't confident enough), Nancy (my best friend who's a child detective), Carlita (my best friend who's a hairdresser), Rachel Carson (my best friend with the Rachel), Maddie (my best friend who was created by the big bang), Lizzy (my best friend who was created by the big bang), Sophie (my best friend who was created by the big bang), Anna (my best friend who was created by the big bang), Robyn (my best friend who takes crazy good pics), Natalie (my best friend who married a cowboy),  Alexandra (my best friend who once bought a painting of Michael Jackson from a homeless man), Francesca

FIG. 0.8

(my best friend who has natural blonde hair), Maggie (my best friend who killed a dolphin), Pilar (my *mejor amiga* who is DEFINITELY a naturalized citizen WINK WINK), Emma (my best friend who evaded her taxes in the most adorable way), Claudia (my best friend who is a kickass New York book agent), Sasha (my best friend whose sister is Malia), Rachna (my best friend who named her son Eli), Rakhee (my best friend who's Rachna's sister), Jen (my best friend who knows her place is in the home), Alyssa (my best friend who has dirty dreadlocks that smell awful, like someone shit in a clown's mouth), Sophia (my best friend who's basically my cousin), Adina (my best friend from Washington State), Ava (my best friend from Washington, DC), Salom (my best friend who's so beautiful even though she's Ethiopian), Hamm (my best friend who is a performer named Hamm Samwich), Isabel (my best friend who can run really fast), Emily (my best friend who grew up across the street from me), Abigail (my best friend who has two really cute kittens named Cocoa and Pebbles), Mia (my best friend with the botched belly-button piercing), Madison (my best friend when I need a designated driver), Elizabeth (my best friend who doesn't know how to read), Avery (my best friend who can speak Tamil), Addison (my best friend who lives in a historical house), Mackenzie (my best friend who is literally a leprechaun), Giana (my best friend who pronounces *mozzarella* like a real Italian),  Faith (my

FIG. 0.9

best friend who is a women's prison correctional officer, you go, girl!), Melanie (my best friend who is an advice columnist), Blanche (my best friend who's literally a white pit bull), Sydney (my best friend who has a really big crush on Jeff Bridges, WTF), Bailey (my best friend who is such a lightweight she can only drink like five drinks a night??), Caroline (my best friend who ate so many carrots her skin turned orange), Naomi (my best friend who is so beautiful even though she's Grenadian), Morgan (my best friend who looks like a Chanel model), Kennedy (my best friend who was an MTV VJ), Lindsay (my best friend when I wake up with stigmata [only happened a couple times]), Audrey (my best friend when I need to practice kissing), Savannah (my best friend who still truly believes in African colonialism, WTF), Sarah (my best friend who likes to take photos of dogs in costumes), Alissa (my best friend who wrote a book on art), Claire (my best friend who's an oil baron), Taylor (my best friend to spit on sidewalks with), Riley (my best friend who doesn't know what "fellatio" means, LOL), FIG. 0.10 Camila (my best friend with a bowl cut),

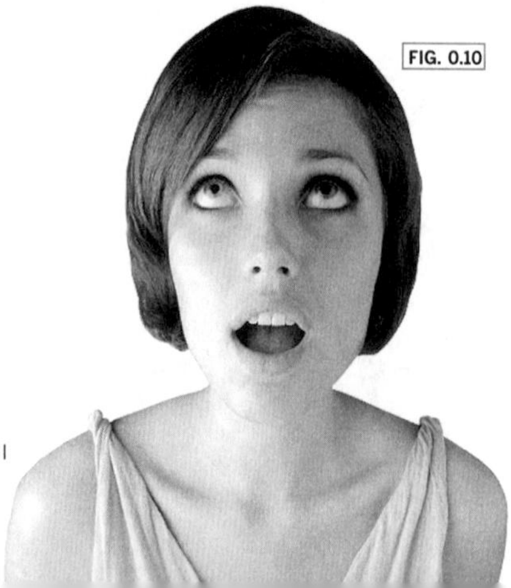

FIG. 0.10

Brianna (my best friend who lives in Frank O'Hara's old apartment), Rheeqrheeq (my best friend with the most silent $q$'s in her name), Peyton (my best friend with two headbands), FIG. 0.11

FIG. 0.11

Bella (my best friend who is not anemic), Meg (my best friend who won a Tony), Alexa (my best friend who is an assistant to an opera singer), Kylie (my best friend who has beautiful dark chocolate-brown skin), Kira (my best friend who's Kylie's super hot little sister), Dereka (my best friend who lied to me once about dating a fifty-year-old), Benita (my best friend when I need to be photographed naked in a bathtub), Max (my best friend who I am literally in love with), Scarlett (my best friend with really awful tan lines), Stella (my best friend with a disgusting beer belly), Maya (my best friend who totaled her first car), Catherine (my best friend with the worst posture), Julia (my best friend who is so into the Space Needle it's weird), Lucy (my best friend who is a photographer but like went to school for it), Madelyn (my best friend when I'm using the Instagram filter "Valencia"), Autumn (my best friend when I need someone to do my makeup really well),  **FIG. 0.12** Summer (my best friend

**FIG. 0.12**

named after a season [sorry Autumn!]), Ellie (my best friend who dated Skrillex), Jasmine (my best friend when I need a massage), Chris (my best friend who 1 percent of me thinks killed JonBenét Ramsey), Skylar (my best friend who is a tiny little bird), Kimberly (my best friend who likes to make props out of stuff), Violet (my best friend who is a color), Molly (my best friend who was in the same sorority as her sister but a few years apart), Drew (my best friend who is half Armenian, half Asian, it's so weird), Aria (my best friend who can't take a fucking joke), Jocelyn (my best friend whose acne is so bad she basically is the Phantom of the Opera), Trinity (my best friend who wears the same tampon size as me), London (my best friend who is covered in a little cloud at all times like Pig-Pen), Lydia (my best friend who's a tiny bug), Annabel (my best friend who is the color of shit), Jessica (my best friend who can eat those huge steaks in under an hour and get them for free), Jennifer (my best friend who dated a guy who worked in our dining hall in college), Jaycee (my best friend who married this guy that I used to have a really big crush on, oy vey), Stephanie Sondheim (my best friend who wrote every piece of art that is important to me and who I love so much I can't even begin to describe or I will cry on my manuscript), Angie (my best friend who totally saved me at my first job), Brittany (my best friend who I can talk to about anything as long as it's boys or apocryphal Anglo-Saxon history texts), Nicole (my best friend who's my personal trainer), **FIG. 0.13** Heather (my best friend who was born homeless and will surely return to poverty soon), Barrett (my best

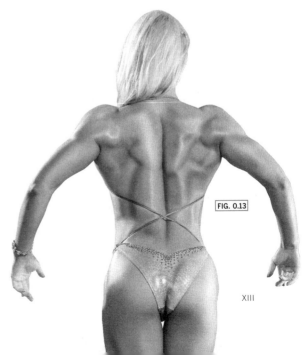

**FIG. 0.13**

friend who looks like Winona Ryder), Megan (my best friend when I want to have a friend with the same name as me), Samantha (my best friend who's married to a literal rocket scientist), Melissa (my best friend who is a dominatrix and can stick a candle up your butt for so cheap no joke), Danielle (my best friend who has this little microphone headset), Amber (my best friend who grew up on a hippie commune), Maxine (my best friend who right now has a candle up her butt), FIG. 0.14 Laureen (my best friend from elementary school whose face got so fat it's insane), Rachel (my best friend who is afraid of taking drugs), Kim (my best friend who got hit by a motorcycle in Calcutta), Laura (my best friend who was in

FIG. 0.14

*The Sound of Music*), FIG. 0.15 Amy (my best friend who's like also a mentor but also she's less cute than me so she's not a fashion mentor), Kayla (my best friend with SARS), Katherine (my best friend when I'm feeling like a little kitty cat, purrrrr!), Sara (my best friend who has the largest boobs

FIG. 0.15

proportional to her frame), Kelly (my best friend who had kids really young), Erica (my best friend who's a baby), Whitney (my best friend whose favorite part of the Holy Trinity is the Holy Ghost, WTF), Courtney (my best friend when I'm wearing a corset), Erin (my best friend even though she has a boy's name), Angela (my best friend who's currently having a heart attack as I write this, hang in there, babe!), Jan (my best friend who's my aunt), Andrea (my best friend who literally is my cousin), Jamie (my best friend who can tell the future and she said I'm going to be so skinny in 2042!!), Lisa (my best friend when we're in upstate New York), Tammy (my best friend to slap because she has such a high pain tolerance), J.J. (my best friend whose name basically looks like a typo), Julie (my best friend who can tie a cherry stem in a knot using only her mouth, hands, and glue), Tracy (my best friend when we're getting acupuncture), Dawn (my best friend to wash oil off little baby ducklings during oil spills with), Karen (my best friend who turned from an ugly duckling into a butterface swan), Susan (my best friend who is deathly allergic to peanuts, buzzkilllll), Christine (my best friend who won't eat anything but buttered noodles), Cynthia (my best friend who used to own her own ballet studio), Lori (my best friend who's a body pillow), Patricia (my best friend who played cello and her best friend was Anna Wu, that bitch), Pamela (my best friend who only made junior varsity in soccer, what a loser), Wendy (my best friend who never wants to grow up), Sandy (my best friend who got thigh lipo ["thigpo"]), Stacy (my best friend who puts cheese on her apple pie), FIG. 0.16 Debbie (my best friend who's an awesome nurse), Nita (my best friend who I used to think was named Pita, so random!), Carolyn (my best friend who looks so great for her age), Bernice (my best friend who is an organ donor [TMI]), Betsy (my best friend who's an anesthesiologist), Janice (my best friend who has a collection of nutcrackers), Shannon (my best friend who is a collection of nutcrackers), Kit (my best friend who ghostwrote the Bible), April (my best friend whose brother was my prom date), Lesley (my best friend who is so glamorous she wears fur sports bras), BenSimone (my best friend who I think might be JonBenét all grown up and in hiding), Lindsey (my best friend who has too few or too many livers, I don't remember), Kristin (my best friend who loves to "mess" with "Texas"), Roberta (my best friend with the most Twitter followers, that bitch), Ezri (my best friend who is so in love with boat shoes, WTF, it's like marry them bitch), Blostam (my best friend whose name was supposed to be Blossom but they made a

FIG. 0.16

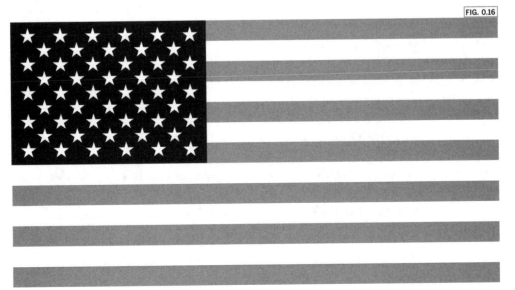

typo on her birth certificate), Edna (my best friend who married my best friend Chad and didn't invite me, so on second thought are we best friends, really?! Answer: YES!), Bradena (my best friend who worked at an amusement park and secretly dated one of the kid visitors for like two weeks), Alicia (my best friend who married her high school sweetheart, TOO CUTE!), Donna (my best friend who works at Dunkin' Donuts, be careful of those donuts, Donna, so many carbs!), Rose (my best friend who's a lesbo #YOLO #nohomo), Petra (my best friend who is this amazing female actress and comedienne), Aparna (my best friend who is an Indian, dot not feather, though I bet she would wear a feather, she has such bad fashion sense, barf!), Vanessa (my best friend whose favorite movie is *Never Been Kissed*), Augusta (my best friend who took Krokodil and now her skin is on the floor), Audra (my best friend who is a goddamn diva), Ronnie (my best friend who sometimes forgets to eat and I am so jealous of that I sometimes pepper her Diet Cokes with those silica gel packets they put in dried foods), Jonna (my best friend who loves the Cheesecake Factory almost as much as I do #CheesecakeFactory #Glamburgers), Artis (my best friend who taught me how to write) FIG. 0.17 Natasa (my best friend from Croatia), Billie (my best friend who yells so much she feels like the HUMAN VERSION OF CAPS LOCK), Ashton (my best friend with mild OCD), Mary Ellen (my best friend when I'm in the "Deep South" [giving

head]), Tricia (my best friend who's a poet statue made out of clay), Kara (my best friend with bushy hair), Mary-Todd (my best friend who got a brain tumor which sucks because they had to shave her head and she had such great bangs), Bridga (my best friend who is fleeing on horseback from her volleyball team), Kiki (my best friend who has the cutest dog Pip), Sammi (my

FIG. 0.17

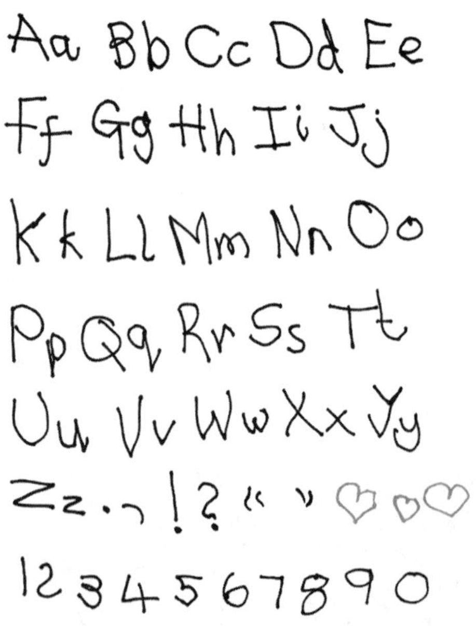

best friend who is an inner-city schoolteacher), Aleks (my best friend when I'm in Richmond, Virginia), Juliet (my best friend who's a fun and flirty place mat), Maria (my best friend who fell off a roof and now can't talk so basically she's the best listener!), Lolo (my best friend who invented erasers I think), Mrs. A (my best friend who taught me first grade), Celine (my best friend who is Céline Dion), Mrs. L (my best friend who makes meat pies), Kathryn (my best friend that's an apple [not apple-shaped, she's a literal apple]), Tara (my best friend who burnt two of her fingers to the stump, weirdly), Magda (my best friend who has a British accent), **FIG. 0.18** Monica (my best friend to suck the president's dick alongside), Jacqueline (my best friend to dip in

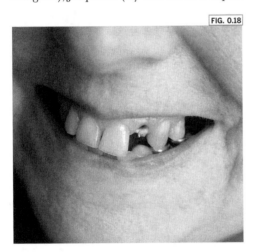

FIG. 0.18

ketchup and wrestle with), Holly (my best friend who was the baby on *Breaking Bad*), Cassandra (my best friend who is a figure from ancient Greece), Brandy (my best friend who's married to the sea), Chelsea (my best friend who has a podcast), Brandie (my best friend who has pink hair), Leslie (my best friend when I want to get

froyo), Diana (my best friend who was literally a candle in the wind), Dana (my best friend with food poisoning), Jenna (my best friend who has a birthmark so big I only look at it and I've never seen her face), Brooke (my best friend whose eyes are literally bigger than her stomach, she is a freak), Matilda (my best friend who is Jackie's roommate), Valerie (my best friend who is a witch and has cast four VERY powerful spells on my reproductive organs), Caitlin (my best friend who is an actual wino), Stacey (my best friend who can sew hats), Brittney (my best friend who stole my health-insurance card but she gave it back), **FIG. 0.19** Margaret (my best friend who is Greek but she's so swarthy you'd think she's Tunisian

FIG. 0.19

or something), Sandra (my best friend who invested a lot in Slim Jim stock), Tali (my best friend who once invited a bunch of black people to a party, AWKWARD), Joanne (my best friend who is I think like an actual British dame but she might just be like a British dock worker with a cool accent), Phyllis (my best friend who has such an old-lady name she should be dead, and she is!), Lucille (my best friend who is my landlady, sorry about the clogged toilet, babe! It was Taco Tuesday!), **FIG. 0.20** Candice (my best friend who won a MacArthur "genius" grant even though

she's basically retarded), Nasia (my best friend to roast bones with), Meghan (my best friend who has my name but she spells it wrong because she's retarded), LaToya (my best friend to speak Yiddish with), Bethany (my best friend to eat the crusts off my pizza for me, she is a cow!), Misty (my best friend who is so stupid that she actually believes horoscopes [makes sense though, she *is* a Leo]), Katrina (my best friend who is classically ugly), Karey (my best friend who has super hipster glasses, they don't even have lenses in them! Side note: Karey is legally blind), Kelsey (my best friend who's not a lesbian she's just super athletic), Joy (my best friend when we're using Photoshop to reduce my thighs into tiny little chopsticks), Jillian (my best friend who has huge cankles), Denise (my best friend who pronounces her name like it rhymes with "Vanessa"), Sabrina (my best friend who worships the devil), Gina (my best friend when I'm hiking up to the Hollywood sign), Jill (my best friend who's a sassy manager), Eryn (my best friend who's a sultry manager), L.W. (my best friend who's a kickass LA agent), Gregoria (my best friend who's a kickass LA agent with stubble like a man), Daniella (my best friend who sends all our friends lemon pound cakes for Christmas every year), Alana (my best friend who tore her Achilles playing basketball like a real idiot), Michaela (my best friend who I will forever be indebted to for giving me my awesome job), Bennie (my best friend who I regret telling that I once hemorrhaged blood into my

FIG. 0.20

shoes), Marina (my best friend whose nickname is Panther), Donica (my best friend who won't give a blowjob unless he dips it in orange juice first),  FIG. 0.21 Sam (my best friend when we're doing the *New York Times* crossword puzzle), Harissa

FIG. 0.21

(my best friend who loves devil's three-ways), Naftalia (my best friend who loves sound effects), Golda (my best friend who has never been tested for AIDS, WTF), Norma (my best friend who's Canadian), Dani (my best friend who hates firemen for a bunch of reasons), Phillipa (my best friend who has a super hot brother), H.P.T. (my best friend to play Hey Lolly with), Jackie (my best friend who knows how to mix all these amazing drinks like margaritas OMG so good), Jane (my best friend who I'm extremely attracted to),

Fluffy (my best friend that I love so, so much despite her extra skin), Branty (my best friend who is a goddamn great editor with a goddamn stupid name), Alex (my best friend who is my twin who I love even though Alex used to call me a Chunk Spelunker to make fun of me for being fat and also Alex made me sign a contract saying I owed Alex my hologram Charizard Pokemon card even though I didn't need to sign it, it was all a ruse, the contract was in crayon for chrissakes), Carol (my best friend who's my mom, mother-daughter relations can be tough but I love her so much but also like it's tough, you know? But honestly she's my best friend and I wouldn't know what to do without her. I owe literally everything in my life to her. But also it's like tough, you know?). FIG. 0.22

And most of all: you. You are my *real* best friend.

FIG. 0.22

# Hello!

Hey ladies!!!!!!!

Welcome to my book. Thank you for coming. I am so overjoyed you were able to make it today. As a busy modern woman in today's modern lifestyle, you could have been anywhere, but instead you have opted to make room in your schedule to be at my book. So: thank you.

And, if you're reading this, congrats are in order: you can read! Literacy is tough in a girl's busy schedule, and I commend you on being able to balance work, family, and reading. You go, girl! Can I get a "what what"?

Let me pause for a moment. This is going to have to be on an honor code, since I will never actually be able to tell if you're giving me said "what what" or not. That's just the implicit agreement that we're going to enter into if you're going to be reading this book with me. When I ask you for things, like "what whats," I'm going to just assume you're audibly giving me them. From now on, I trust you. Ladies: don't lie to other ladies, ladies. If I feel like this isn't working, I might have to start making you mail in audio recordings of you giving me my rightful double whats, and none of us wants that.

Now, on to why you're here. Science. From the Latin *scientia*, meaning "science." Science is the study of the beautiful world around us. A man looking closely at a woman's boobs? That's science (anthropology). Ted Bundy cutting off a woman's boobs so he can wear them? Also science (erotic anatomy, or "manthropology"). Anthropologie? A cute store!!

My book *Science . . . for Her!* is absolutely vital for the female population. (NOTE: I was going to call this book *Science for Dummies*, but apparently that was already taken.) Science is hard for most people, let alone *women*. It has been demonstrated repeatedly throughout history: female brains aren't biologically constructed to understand scientific concepts, and tiny female hands aren't constructed to turn most textbooks' large, heavy covers. **FIG. 0.23** In school, I suffered through the pain of having to learn from math and

**FIG. 0.23**

3X

typical woman's hand

typical man's hand

science textbooks designed for men. Anatomy?! More like I-Can't-Mommy! The subjects were incomprehensible. I was so bad at math that I became a writer. And even that was hard. Writing is like word-math! In this book, I present basic high school math and science concepts in a way that is tailored for the female brain (in the form of a *Cosmopolitan* magazine!). You'll instantly be set at ease by the way my book mimics a women's-interest periodical, your comfort zone. Fun quizzes!

Sex advice! Skorts! The page numbers may be out of order (since math is VERY hard), but the content is all there.

My life used to be completely different. I didn't know a thing about science, and I *couldn't have been happier*. I would go through my days oblivious to how ignorant I was, indulging in my favorite hobbies of designing quinoa-inspired sunglasses and buying Lasik for my dog. But one day my then-boyfriend Xander told me we had to talk. He sternly sat me down on his face and told me we were breaking up. I started sobbing and begging—I needed him! What did I do to deserve this! He calmly explained it was because I didn't know any science. He was upset that I had lied on my résumé and was now in charge of NASA's northwest division and responsible for many, if not all, previous space shuttle crashes.

That was very eye-opening for me. That very day, I bought myself a bunch of science textbooks and the cutest purple bra and three pints of Ben & Jerry's and humped two guys named Ben and one named Jerry. I worked harder than I ever had before. In the very first day, I was able to teach myself chemistry, geology, and how to slash a car's tires. **FIG. 0.24** On an unrelated note, my ex-boyfriend Xander's tires got slashed. Can I get a "yee-haw!"? I assume I did!

Most of you, like me, have probably experienced the supreme embarrassment that comes from not knowing the science that men have mastered at an early age. Up until now, the only female science textbook was *Men Are from Mars, Women Are from Venus*, and honestly, I've recently learned that that's a metaphor and not an advanced astrobiology textbook. Score one for you guys, former NASA coworkers! You were right! I have no use for books that feed impressionable women a load of metaphoric language. Like *Where's Waldo?* No one can ever actually *find* Waldo. It's just a stupid metaphor about how we're all metaphysically adrift in a harsh world that feels like a homogeneous sea and OH MY GOD THERE HE IS! On the rock behind the big top of the circus!

**FIG. 0.24**

(NOTE: I apologize, I was doing a *Where's Waldo?* book during the writing of this introduction. I promise that won't happen again. Can I get an "I accept your apology, Megan"?)

Read this book, and you, woman, will become a master of your surroundings. Here are just a few things you'll learn how to do through science!:

• Make a biological clock out of a potato!
• Learn scientifically why women can't drive!
• See which microbe on Mars is the best kind to date!
• Slash your ex-boyfriend's tires!
• Take a fun, flirty quiz to see if you have cancer!
• Good riddance to him, he sucked!
• Memorize physics formulae through nail art!
• Use chemistry in the kitchen!
• His face was super cute, though!
• In fact, he kind of looked like Waldo!
• Maybe that's the reason I need to "find" Waldo, he looks like my ex!
• Or maybe it's because I never had a dad!
• I love you, Daddy-Waldo!

I also promise that, to help with the brain pains that will come with so much learning, I will check in every once in a while and let all you ladies know the latest fashion trends that I'm wearing. Right now, I'm wearing a Juicy Couture tracksuit top, a Louis Vuitton pantsuit bottom, and a Victoria's Secret buttsuit (that's what I call underwear).

Though it might not seem like it now, this book is going to be the most important thing in your life pretty soon. More important than air, blood, and fun lotions combined (ew, don't combine those!). Not to be a buzzkill, but it's our responsibility as modern women to take control of our lives and academic relationships with men, and learning science is the first step to an equal intellectual footing for gender equality. Oh, speaking of steps and footing, I almost forgot: I'm also wearing Prada footsuits (shoes)!

Please just put your faith in me, babes. As a self-professed strong, successful, educated, bloated (ugh!) woman, I am the perfect person to attack an institution of gender inequality in the sciences. And everyone loves when a woman gets on a soapbox! You can see up her skirt!

Thank you very much for getting through my introduction. In the immortal words of Isaac Newton: "Please enjoy the rest of Megan Amram's book. I'm Isaac Newton." Now set your biological clocks to study time, because here we go!

MEGAN AMRAM
*Writer / Alleged "Astronaut Killer"*

# Letters to the Editor

"This is
SO
great!!!"

—my best friend Carly

"Please cease and desist. This is not sound science and could be gravely detrimental and dangerous to the female population. If you do not recall these books from the shelves, we will threaten legal action."

—National Science Foundation

"I love you
SO MUCH,
Carly!!!!!"

—Me!

# About the Author

# Megan Amram

is a fun, flirty young woman living in Los Angeles, California. Credentials include a BA in medicinal fashion from DeVry and a BMI that was in the "unhealthily underweight" range for one day after a stomach flu. She dated her "ex"-boyfriend Xander for three years before he broke up with her, but she's pretty sure they're gonna get back together pretty much ASAP because they have each other's names tattooed on their wrists so legally they have to be together, it's a double-jeopardy thing. Before working at NASA she was funemployed (unemployed in a cute jumper).

# About the Author's Ex-Boyfriend

# Xander Mince

is a five-foot eleven, green-eyed, rich-businessman love of Megan's life. One of his
eyes is a little wonky but whatevs, his dick is basically the size and girth of
an ocean buoy so his ex Megan is not complaining. He's great at sex and his dong
was literally a million miles long. After he and Megan broke up a couple
weeks ago due to this so-called NASA "debacle" he told Megan he was going to
move somewhere far away enough that he couldn't smell her "stench of ignorance,"
but Megan knows he's still living in Los Angeles because she drives by
his house a lot and honks until he threatens to call the cops but they'd never
catch me, I mean Megan.

# Biology

# Introduction

OMG, we're really doing it—the science part!! Oh my G, this is just too fun! I knew this was a fun project, but I had no idea it would be this fun, and *easy*. I'm taking a selfie right now of me writing the first few real lines of my book!!  FIG. 1.1

FIG. 1.1

If only my ex-boyfriend Xander could see me *actually writing a science textbook*, he would just *die*. You know what, I'm going to text him this photo. I deleted his number after we "broke up," but I had memorized it. Trick: it's easy to memorize a ten-digit number when you have it tattooed on your knuckles! Who's got the last laugh now, Xand?!

Just want to stop and check in for a second—how am I doing? In terms of the science book, but also in terms of our friendship. Do you like me? I've always gotten along better with boys, so it's a little hard for me to make friends with women sometimes. I'm not one of those girls who's into drama and gossip. Like my best friend Miranda, who, in all confidence (DO NOT TELL A SOUL), just had her *seventh* miscarriage (though everyone knows she's lying, let's just say she has been "pro-choice" seven times). I just don't relate to gossipy girls like that!

Okay. Let's *do this*. Biology comes from the Greek *bios-logia*, which, loosely translated, means "biology."

. . . YES! We're IN IT NOW, GALS! You know what else comes from Greece? Feta cheese and hot guys. I spring breaked (spring broke??) in Mykonos in 2009. And let's just say Theoros left me with a little souvenir feta cheese (yeast infection). Science!

Biology is the study of life. Life is all around you if you haven't noticed, stupid-dummy! I'm sorry—that was woman-on-woman hate, and I do not condone that. It just slipped out of me. I must be jealous of you. You do have nice bangs. I truly apologize. :( (You are my best friend.) (Bitch.) (What is going on with me? I am the bitch. I must have PMS.) Which segues nicely to the next section . . .

# Reproduction

I'm obviously going to start a science textbook for women with the most important science there is: reproduction. Let's see . . . how can I put this in a way that you ladies will understand . . . PUTTING THE HOT DOG IN A BUN. DRIVING THE TRAIN THROUGH THE TUNNEL. BATTERING THE PARSNIP AT THE STATE FAIR. BURYING THE BONE-IN RIBEYE IN A TRIBECA GRAVEYARD. RIDING THE SKIN BUS TO TUNA STREET, MASSACHU-TITS. CHECKING THE PUSSY OIL IN THE VOLVO CROSSCOUNTRY. GLAZING THE DONUT WITH SPERM-ANENT GLUE. DEEP DICK SOUP. DOING THE MOMMY-DADDY GANGNAM STYLE.

. . . Gals, if you haven't figured it out already: I'm talking peen in veen! You know: SCREWING THE LIPSTICK BACK INTO THE LIPSTICK TUBE. FIG. 1.2 TAKING THE ONE-EYED VINCENT VEIN GOGH TO THE OPTOMETRIST. DECLARING A WIENER AS A DEPENDENT ON YOUR TAX RETURN.

Sorry, I can barely get through this without making a double-entendre-style euphemism! I'm talking let's talk about sex, baby. Those are two different topics: "sex" and "baby." We'll start with sex. Sex begins when you let a boy—or a man, because come on, girls, hold yourselves to a higher standard!—put a wiener in your body. In humans, once the corn dog (or "penis") has been battered (or "wiggled around and exploded like a snail in a microwave"), a tiny nugget of a baby is planted in the woman's uterus (or "uterus"). Sperm, that stuff that normally goes on your chest or face, can go into your vagina ("Hurt Locker") and combine with eggs that are inside of you. Yes, you have millions of eggs inside of you at all times—many of them human! Unless you're old and have gone through menopause, and if that's the case, what are you doing reading this book? You should be spending the golden dusk of your life marrying your dressmaker dummy and bedazzling your AARP card! Bu-BYE, OLD MAID! Don't let the door hit your sagging ass on the way out!

FIG. 1.2

Only one sperm can fertilize one egg and that makes a baby. Babies are then constructed piece by piece. First their arms (two, if possible! It's easier to wear blouses that way!). Then legs (same, aiming for two, pants usually need two, skorts are a bit more forgiving). A taut little stomach (let's hope your baby loses that baby weight quickly!). Then the other body parts that we'll fully discuss in the ANATOMY chapter. Then your baby gets a tiny wiener ("baby carrot") or tiny v-eener ("mini-clam piñata"). But before any of that can happen, the baby has to be started through the fertilization process. FIG. 1.3

FIG. 1.3

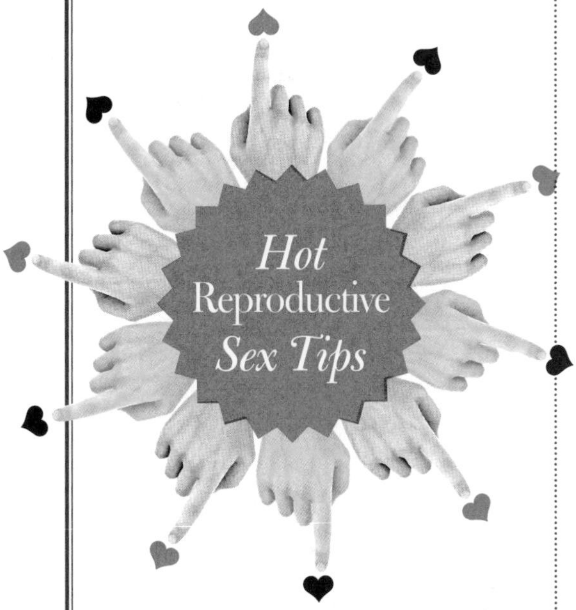

*Hot Reproductive Sex Tips*

**Here are some fun ways to spice up fertilization! Utilize the female body to make your next bout of sex with your guy absolutely mind-blowing!**

**1.**
When he's close to the moment, surprise him by releasing an ovum into your fallopian tubes!

**2.**
The next time you're on top, try attaching the embryo to your uterine wall. You'll be moaning in near agony (in nine months)!

**3.**
At the exact moment that a sperm enters the wall of the ovum and meiosis begins, stick your finger in his butthole!

**4.**
As soon as you feel the *glans* of the penis skim the front of the *cervix* as he exits you after *fertilization*, stick your finger in his butthole!

**5.**
To surprise your man while he's making you eggs the next morning, go up behind him while he's cooking in the kitchen and stick your finger in his butthole! While it hasn't been definitively proven to affect fertility, it's still a great place to put your finger!

If you're not ready to have children yet, *birth control* is the means by which you can prevent unwanted pregnancies. There are many types of birth control: *condoms, birth control pills, being an ugly bitch,* and the most controversial of all: *abortion.* Gals, to be frank, I'm pro-choice. It just doesn't make sense to me that a bundle of cells only a few days/months/years old is really a *baby*. Let's face it, people are just cells until they're eighteen, and even then they just become cells with a dick attached. Though if life begins at conception, then I can use the carpool lane for the next few days. I had quite a big weekend with a rugby team! FIG. 1.4

Don't feel bad if you hate birth control. It's perfectly normal to not like using birth control. You can't feel anything when you use a condom! It's basically like being drugged, without any of the good effects of being drugged! Best part of

waking up from a date-rape drug? It definitely WAS a date! No guessing what those texts mean anymore! ;) FIG. 1.5

FIG. 1.5

But birth control can be fun, girls! Don't think it has to be all stuffy! There are plenty of fun, flirty ways to keep that baby nonexistent for the time being!

FIG. 1.4

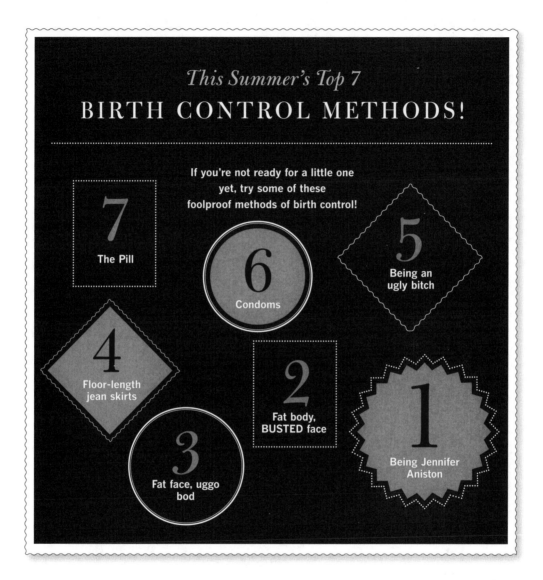

## This Summer's Top 7
# BIRTH CONTROL METHODS!

If you're not ready for a little one yet, try some of these foolproof methods of birth control!

**7** The Pill

**6** Condoms

**5** Being an ugly bitch

**4** Floor-length jean skirts

**2** Fat body, BUSTED face

**3** Fat face, uggo bod

**1** Being Jennifer Aniston

One day you might be thinking, "I can't have children! I can barely care for myself/my dogs/my two kids!" Then the next day, you're a shivering wreck, swaddling a yam in baby clothes and singing it to sleep and robbing a sperm bank and sitting on all the full sample cups that you just robbed. That's your biological clock going off.

A woman's biological clock is just ticking within her (scientists have narrowed down the location within the body to somewhere around the aorta).

Let's put this complicated biological-clock business into lady-mag terms you can understand. Here's a fun recipe for how to build a biological clock out of a potato and some other simple ingredients!

# HOW TO BUILD A BIOLOGICAL CLOCK OUT OF A POTATO

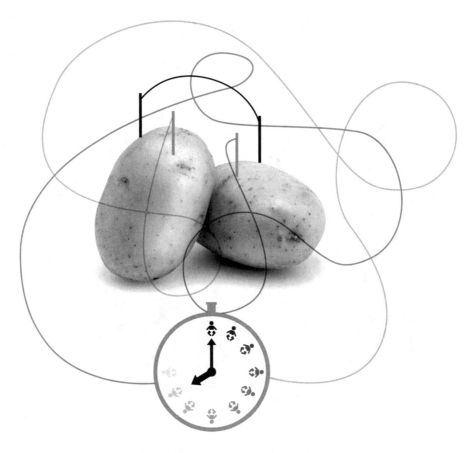

Literally every woman has a ticking time bomb within them that someday will explode and tell you to have millions of babies like a disgusting spider striped with stretch marks laying a fetid egg sac. This is called a "biological clock." It might explode when you're sixteen or it might explode when you're forty-one (Old Maid much? Didn't I tell you to leave?! GET OUT NOW! SHOO, BITCH!), but it's going to happen.

Did you ever make a potato clock in sixth grade? Well, in a variation on the traditional scientific experiment, here are the instructions to make a biological potato clock!

**10**

Make him make one for you.

**11**

Now's the part where you set the time. It's a little subjective—you're going to have to feel out deep within yourself when you think the all-encompassing urge to have children is going to take full control of your life and mind and body. One thing to look out for is that maybe it's already happened and the urge you're feeling within you is a six-month-old fetus. Once you've decided on a year and date and time for your biological clock, have your husband set the clock face that he put on your clock when he made it for you.

*Note: If you were assaulted by your husband, follow step 12. If you were not, skip to step 13.*

**12**

No worries—if it's your husband, it doesn't really count! Feel free to shower!

**13**

Congrats! You've made your biological potato clock! You go, girl!!!!!

---

**1**

Buy two potatoes on your normal trip to the grocery store. Make sure to ask your husband/lover if he'd like potatoes as well, since if he wants potatoes you should make sure to get enough to make a clock and to give him his potatoes! Yum!

**2**

Also ask your man if he'd like any other groceries while you're there for the potatoes. There's a good chance he'd like something! Men are very hungry! For example, they sometimes like pita bread!

**3**

Get in your car. Make sure to adjust the side mirrors to your level if it's your husband/father's car—they are taller than us! Remember to keep your eyes on the road and NOT to readjust your makeup in the mirrors as you drive. Women need to be extra vigilant while behind the wheel due to their inherent lack of spatial reasoning. Plus, your makeup already looks great, girl!

**4**

Pull over. You got lost on the way to the grocery store, didn't you? Just ask a kindly gas-station attendant or barkeep for directions. They'll be sure to help you if you give a little smile and show a little gam!

*Note: If you were assaulted by a gas-station attendant, follow step 5. If you were not, skip to step 6.*

---

**5**

Do not shower. Immediately call 911.

**6**

You're at the grocery store! Isn't this fun? Head to the potato section. Find two potatoes that are particularly large and starchy. If you aren't sure which potatoes to get, ask one of the greengrocers for help. He'll love the chance to be of service!

*Note: If you were assaulted by a greengrocer follow step 7. If you were not, skip to step 8.*

**7**

Do not shower. Immediately call 911 after you buy the potatoes that he recommends (he still knows his stuff, after all).

**8**

Buy the potatoes. Drive home—carefully! Strap the potatoes into the passenger seat as a precaution, lest you get in an accident on the way home due to your poor spatial reasoning. Don't become too attached, though—they are NOT your babies! They are potatoes!

**9**

Have your husband/father/milkman google "how to make a potato clock" online. It involves nails, wires, etc. Very complicated.

# How Long Should You Wait?

How long should you wait after meeting someone to have sex? It's one of history's greatest unsolved mysteries, right up with the Bermuda Triangle and where Waldo is (lost him again). But I finally have an answer for you. How long should you wait? No more than ten minutes after meeting someone. Preferably under seven. FIG. 1.6

This is not the advice you'd expect from a woman, necessarily. It's widely believed that women tell each other to wait months, even weeks, to have sex with a man. This is SABO-TAGE. Women are giving this piece of "advice"

FIG. 1.6

1 min:
**Marry him now!**

4 mins:
**Start writing your vows**

7 mins:
**Ideal timing**

10 mins:
**Absolute cutoff**

15 mins:
**He must be gay**

to each other so that they can swoop in and steal that man while their reluctant and gullible friend is waiting abstinently. One woman telling another woman to "wait" is a very smart evolutionary tactic to become impregnated with the seed, or "not in my hair," of the strongest male of the tribe. This has been proven incontrovertibly by science, or "number-words."

Back when men and women were cavemen who lived in caves, women were the more aggressive sex and would often fish men's condoms out of the cave-trash to steal their seed. FIG. 1.7 This later evolved into a more passive-aggressive approach where cavewomen would have cave margaritas and tell their girlfriends to respect themselves, and then, when their cavefriends were in another part of the cave blending up the ice for the next round of 'ritas in the VitaMix, the savvy cavewoman would stay behind and have lots of hot caveboyfriend-stealing sex, the moans and grunts of which were conveniently drowned out by the loud blender sound.

FIG. 1.7

Now, in the spirit of full disclosure, I would like to divulge that my best friend Tiffany basically started boning my boyfriend Xander right after we'd broken up while the three of us were spelunking in the Lava Tube Caves in Bend, Oregon. That has nothing to do with this scientifically based truth-based essay and never will!

When someone is perfect, you do NOT want to mess it up by having sex at the wrong time and letting your best girlfriend get in the way and have your baby. That's why you should have sex as soon as possible, preferably within the first ten minutes of meeting someone. If you can do it in under seven minutes post-meeting, you're probably going to marry the guy. Look at me! Xander and I waited eight dates before we had sex, and now he and my ex–best friend are probably gonna get married in their favorite lava tube with a molten chocolate lava cake and bridesmaids that are dressed like virgin sacrifices to a volcano god of yore. It's actually a really cute theme, but still, fuck them.

Now, listen. I can only speak from a woman's perspective. I've always been terrible at impressions, and I couldn't do one of a man if my life depended on it. Actually, that's a lie. I do a great impression of WALL-E, and technically I think he's a man. Here, I'll write a little bit of the impression for you: WAAAALLLLLLL-E. Impressions don't work as well written out!

For most people, the ten-minute rule might seem extreme. How do you know that they're tested for STDs? That they're not crazy? That their pubes aren't shaved into a swastika or a strawberry (Hitler's favorite berry)? **FIG. 1.8** You're going to just have to trust your gut. Or, more specifically, your devil's-gut (vagina). Humans have an evolutionarily evolved ability to know whether their potential partner has crabs or not. Have you ever been drinking twenty beers and about to hook up with someone and then you throw up? That's your crab-sense saving you from disaster.

The actual particulars of having sex within ten minutes of meeting a potential baby-father or -mother are a little difficult and nuanced. The whole situation is touch-and-go. Meaning you touch them and then you go. Let's say

**FIG. 1.8**

you meet a hot guy in a Jamba Juice. He's everything you want in a guy: handsome, correct number of legs, super fuckin' into Jamba Juice. How do you let him know that you guys are going to go do it in the bathroom? I prefer the body-language approach. A flirty wink means "I like you," and saying, "Let's go have sex in the bathroom" means "Let's go have sex in the bathroom."

I'm going to play devil's advocate here for a second. Maybe you should wait until you feel comfortable, or have more than a fleeting physical attraction to the person. HAHAHA, BOO! First of all, that's what the devil's lawyer is saying. You think I'm going to believe him? Sounds like a real snake. **FIG. 1.9**

I hope this helps, dear readers. Now excuse me, I have to go have sex with the person who watched my computer at Starbucks while I went to the bathroom!

**FIG. 1.9**

# To Babe or Not to Babe?

Insanely, some women don't want to have children. I recruited one of my best friends to write a little across-the-aisle piece on why *not* to have children. She's like a second-tier best friend: not close enough that she'd give me a kidney, but close enough that I could steal her kidney.

## NO BABIES, PLEASE!

Hi, everyone! I'm Megan's friend. We met in a laundromat once when she was trying to steal my kidney! I'm here to say: no babies, please!

Most girls have the same life goals: date a boy, get voted homecoming queen (popular and electoral votes), get married, take a picture of a Chupacabra, renew your vows, get divorced, renew your divorce vows, eat a pie behind a middle school, get remarried, and have a baby. Call me crazy, but no baby for me, please! I want my life to be fun and easy, not, as Shakespeare might say, "done and queasy" (SOURCE: *Oxford English Rhyming Dictionary*).

Pregnancy is just a mess. It's like you're a turducken: a woman stuffed with a fetus stuffed with the turducken that you eat every day for breakfast. Your clothes stop fitting, and you have to start buying pants/quinceañera dresses/quinceañera tiaras with elastic waists. You have to start eating and mainlining for two. Sometimes you can't help but sample the cocoa butter that you're putting on your stretch marks. And I want to keep my figure! (For those of you who haven't met me, I'm five foot ten, 120 pounds, 34DDD, my name is Heidi Klum, and I'm the model Heidi Klum.)

Have you ever seen a baby? Or, if you're blind, have you ever touched a baby's face and smelled a baby's face and used echolocation to tell what color it is? They are crazy nasty looking. Also, they're passive-aggressive and love to give the silent treatment. Also, they always try to out-pants-poop me in a pants-pooping contest (as of November 4, 2014, I'm still undefeated).

Some more fun facts: an average baby is approximately six metres long and thirty-one fluid ounces in metric circumference; in comparison, a normal vagina is at most one kilolitre in diametre (SOURCE: British Association of Metric Measurements and Also Obstetrics). Pushing a baby out of your body is like pushing a watermelon through your vagina, and, trust me, that was not a fun Cancun Spring Break 2004 drinking game. The only thing I want coming out of my body is a contented sigh when I've eaten an extra-tasty Toblerone in my baby-free bachelorette pad filled with non-baby-proofed coffee-table corners and sharp Toblerone vending machines.

How about the money issues? I can't afford a child, let alone a kid. With modern science, babies will soon live to be one hundred years old and will grow to be thirty feet tall. Do you realize how much it costs to buy baby food for one hundred years? Diapers alone are thousands of dollars each, if you prescribe to the old wives' tale that you should only use Gutenberg Bible pages as diapers. I need all the money I can get for adult things like coffins and tax-themed Mad Libs. People with babies don't get to be adults anymore. I would hate to give up my right to my height-restricted dinner parties. Call me crazy and Heidi Klum, but I just don't think it's worth it.

Sure, sometimes I get bored and lonely without a baby. There are only so many times you can stage an intervention for your blow-up sex-doll gal pal, even though she really needs to know that she doesn't have to sleep with guys just to feel pretty. And there are only so many times you can buy three blow-up sex dolls and pretend to be *Sex and the City*. But even though my biological clock might be saying, "Have a baby," my biological cell phone voicemail message is saying, "Enjoy your twenties and don't have a baby," and my biological fridge is saying, "Eat that cottage cheese, it's still good."

So, trust me: babies are the worst. No babies, no staying up all night. No having to share the strained peas and mashed pears you've always enjoyed solo. No having to give them fake names like Apple or Anderson Cooper. Always remember, babies are for the weak. And listen to my biological mouth when it says, "My name is Heidi Klum."

*Heidi Klum*

# Rebuttal

You should have babies! FIG. 1.10

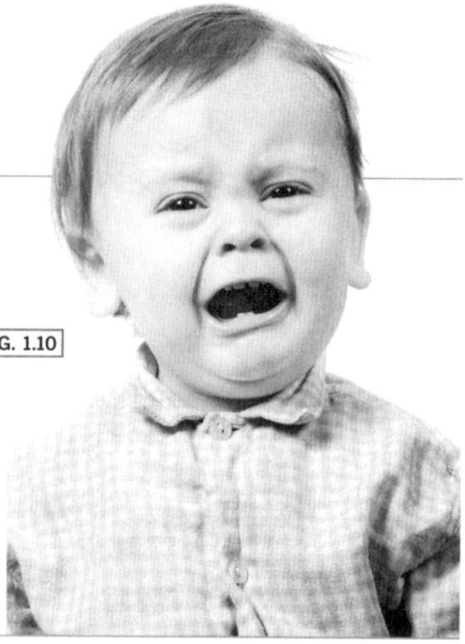

FIG. 1.10

# Cell-Fasteem!

Cells are the basic unit of structure in every living thing, from your mom Michelle to my aunt Michelle (OMG, babe—we're cousins!). Cells were first discovered by Robert Hooke in 1665, who looked at cork under a microscope and thought that it looked like it was made of cute little bubbles, how fun, what girl doesn't like a lavender bubble bath now and again! Then he took the bottle of chardonnay that the cork came out of and poured himself a chilled glass of white wine, which led later scientists to believe he was a homosexual. Ladies, a tip: if your man drinks white wine, he's probably gay. If he drinks red wine, on the other hand, he's probably . . . gay. All men are gay! Ugh! FIG. 1.11

*Cells* are sacs of fluid that are reinforced by proteins. They're held together with *membranes*. Without cells, you would just break apart in a gush of chemicals that would just goosh onto the floor and then no one would ever ask you to prom or to get married because you'd just be a smush of chemicals that just blooshed everywhere and you'd have no boobs to grab onto. Basically you'd look like Kristina did after she watched *Nights in Rodanthe* the first time!! OMG what a mess was she! Kristina has been one of my best friends for crazy long. I later found out that her mom had just found a malignant melanoma (see CANCER section) on her upper thigh so that was kinda more why she was crying a lot. I kept telling her that if you get skinny enough you won't even HAVE an upper thigh to get cancer on but that just seemed to make her cry more. Anyway, I think Kristina liked the movie!

## Featured Sex Moves

SIDE NOTE: "Knights in Rodanthe" is this chapter's featured *Science . . . for Her!* sex move! We will be highlighting one move per chapter for you ladies to try to keep your love (or *lust*!) lives aflame! Knights in Rodanthe is where six or more men with swords bang you at Medieval Times after you change your name to Rodanthe.

FIG. 1.11

# *What Your*
# MAN'S DRINK SAYS ABOUT HIM

If he drinks:
**WHITE WINE**
He is:
**GAY**

If he drinks:
**COSMOPOLITANS**
He is:
**GAY**

If he drinks:
**RED WINE**
He is:
**GAY**

If he drinks:
**MARTINIS**
He is:
**GAY**
(Olives come in salads,
and salads = VERY GAY.)

If he drinks:
**SCOTCH & SODA**
He is:
**GAY**
(Men wear skirts
in Scotland.)

If he drinks:
**MALE SEMEN**
He is:
**GAY**
(Semen contains sperm that will end up being human males at
some point after fertilization and therefore they have little tiny
dicks, so when your man is drinking male semen he's actually
sucking a million tiny dicks. This is almost as gay as salad.)

Nothing's smaller than cells! SUH-SUH-SUH-CELLLSSSSS WICKA WICKA BOW-WWWWWW B-B-B-BOWWWW! That's a dubstep song about cells I just made up for you chic chicks. ☺ FIG. 1.12 Cells are made up of things that are smaller than cells, but those things can't function on their own, so technically a cell is still the smallest part of life. There are two main types of cells. *Eukaryotes* are cells that have nuclei (a *nucleus* is the center of a cell) with tiny organs within them. Eukaryotes include plants, animals, and fungi. Conversely, *prokaryotes* don't have nuclei and are the lowest level of cells. These include bacteria and EX-BOYFRIENDS!!!! FIG. 1.13

FIG. 1.12

FIG. 1.13

## *Differences between* VIRUSES *and* BACTERIA *and* EX-BOYFRIENDS

Viruses and bacteria and ex-boyfriends can be easily mixed up. Viruses are simpler than bacteria and aren't made of cells. They take over other cells to reproduce. Bacteria are fully prokaryotic cells. Ex-boyfriends are similar to viruses in that they take over other organisms and use their husks for their own means until the original host organism is moribund and dying. Ex-boyfriends also resemble bacteria in that they cause syphilis.

## Can you recognize these stars from their
# HIGH SCHOOL PHOTOS?

**Stars had to go to high school, just like us! See if you can recognize these famous celebrities from their early days!**

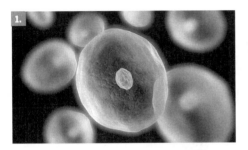

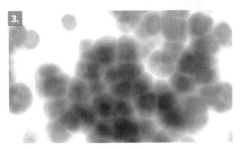

## ANSWERS

**1.**
**Angelina Jolie!**
Angie is barely recognizable as a tenth grader at Beverly Hills High School! Good thing this ugly duckling turned into a swan!

**2.**
**Daniel Day-Lewis!**
Even this young, you can recognize the future talent in his expressive features.

**3.**
**Cameron Diaz!**
Not pointing any fingers, but it looks like *someone's* had a little work done since then . . .

**4.**
**Ted Kaczynski!**
It's very disturbing to look at this early picture of a killer and truly see the evil radiating from within as early as high school.

# Genes

Since it's distracting to bring up a homophone of one of women's favorite things without showing them, I will start off this section with a treat—a glossy page chock full of photos of beautiful jeans!

Phew! I'm glad we got that out of the way! Jesus, I've got the jean sweats! I'm covered in the sour nectar of the sweats that I get when I look at great jeans. I smell like a 7-Eleven meatball sub filled with rotten Starburst.

*Genetics* has to do with how living things inherit traits from their parents. This transmission happens through *genes*. Genes use *DNA* (short for *deoxyribonucleic acid*, what a mouthful! And we all know what to do with mouthfuls, right, ladies? That's right—SWALLOW!!) to transmit traits. For example, my mom is, like, totally amazing and, through DNA, I totally inherited all the best parts of her. **FIG. 1.14**

When a cell starts to replicate, the DNA strand (which is made of two strands curled around each other in a shape called a *double helix*) unzips. And if that DNA strand gains weight over the winter, maybe it will never be able to *rezip* again! That always happens to me over the holidays! I mean, there's cookies, eggnog, Communion wafers—how am I not going to gain weight? Body of Christ? More like Body of *Carb-st*! (NOTE: I am a Jew, but I often attend Catholic Mass services to closet-eat, since none of my liberal Jew gal pals will ever catch me. Only

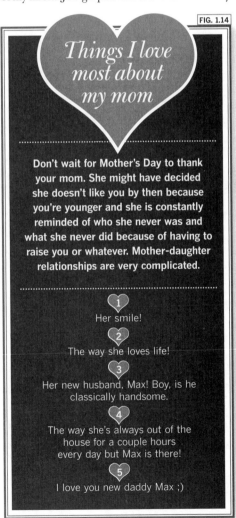

**FIG. 1.14**

# Things I love most about my mom

Don't wait for Mother's Day to thank your mom. She might have decided she doesn't like you by then because you're younger and she is constantly reminded of who she never was and what she never did because of having to raise you or whatever. Mother-daughter relationships are very complicated.

1 Her smile!

2 The way she loves life!

3 Her new husband, Max! Boy, is he classically handsome.

4 The way she's always out of the house for a couple hours every day but Max is there!

5 I love you new daddy Max ;)

my Catholic best friend, Mary Katherine, who loves 2 pray with/for me. I cannot get enough of those wafers. FIG. 1.15 GODDAMMIT! Why do you think Mary stayed a virgin? Because she was so fat from those fucking wafers! Mary, you virgin bitch!!)

---

**NUTRITION FACTS**                     FIG. 1.15
Serving Size: 1 wafer (50g/2oz)
Servings Per Container: 1

| Amount Per Serving | |
| --- | --- |
| Calories 10,000 Calories From Fat 4,355 | |

| | % Daily Value |
| --- | --- |
| Total Fat 50g | 98% |
| Saturated Fat 0g | 0% |
| Trans Fat 0g | |
| Cholesterol 0mg | 0% |
| Sodium 2,080mg | 130% |
| Total Carbohydrate 68g | 3% |
| Dietary Fiber 3g | 1% |
| Sugars 0g | |
| Protein 0g | |

| Vitamin A 0% | Vitamin C 0% |
| --- | --- |
| Calcium 0% | Iron 0% |

---

You know how sometimes you buy a ton of stuff at Costco? Well, when your parents' DNA mixes, it's combined through *chromosomes*. Chromosomes are large bundles of DNA and genes all clumped together, like a lot of toilet paper at Costco. If your 'giner was the size of .000001 of the head of a pin, you could use a chromosome as a dildo. Sperm and eggs (you know—TADPOLES AND LILY PADS. DOTS AND DASHES IN MORSE CODE. SQUIGGLY EYEBROWS AND DIPPIN' DOTS) only have half of the necessary chromosomes for human cells. When they do their sexy dance, they combine all the chromosomes you need. You get some from your mom and some from your dad, which is why you are a mixture of traits from your two parents. Like, for me, I get my smile from my mom and my vestigial penis from my dad. I thought it was a twin until I was like twenty-three! I'm planning on getting it removed next year. You gotta treat yourself sometimes. :) FIG. 1.16

Your parents' genes combining are why there are no other people exactly like you. Unless you're an identical twin, which happens when an egg fertilized by a sperm splits into two fetuses in the womb. You might call these babies . . . *wombmates!* Again, I thought I had a twin until like two years ago. I always felt like she was the "by the books" one, and I was the "fun and flirty" one. We had a ton of joint birthday parties. It honestly was fairly traumatic when I found out she was a vestigial penis. Which is why she has to go. *NOW.*

FIG. 1.16

# FACE MASH

Genes play a great role in the art of choosing a spouse to mate with. Using face-melding software, I will show photos of what my baby would look like with multiple celebrity fathers.

Now, here's a mash of me and my ex, Xander. You know, just a gentle hint about how we should get back together and have a million babies!

One of his eyes is not very symmetrically placed, so if something looks weird, that's it. But his *personality* is symmetrical, and that's all that matters! And his dick. It's the length and girth of a two-liter bottle of Coke Zero.

# Human Life Cycle

Now, no one likes to think about people dying. Thinking about dying makes my skin break out and my herpes flare up j/k I don't have herpes I'm just big-boned! It's so sad to know that literally everyone you know will die someday. FIG. 1.17 SPOILER ALERT!! Your aunt? Dead. That person you ran over with your Vespa accidentally? Dead. (I checked, he died a few days later.) The person who saw you run over that other person? Not dead, as long as he continues to be smart and keep his pretty mouth shut. But yeah, then eventually dead.

Okay, in the spirit of full disclosure, I feel like I should delve more into my history with bodies. This is a part of my childhood that I've never really told anyone before. I found a body down by the tracks once. It was one of the scariest moments of my life. Oh God . . . I'm shaking just writing this . . .

Here's the other thing: it was alive. It was an alive body. It was the conductor. He was gorgeous, but not rich. That was the super scary part. Conductors don't make a ton of money. There aren't even tips on trains. FIG. 1.18 I knew he wouldn't be able to buy me Prada lunch boxes, and I was really into Prada meal totes in those days. We dated for a few months, but my family didn't approve. I finally had to dump him. We were both so upset about it, plus I dumped him

FIG. 1.17

## WHO'S GONNA DIE?

Your aunt? → DEAD

Your dog? → DEAD

Guy hit by Vespa? → DEAD

FIG. 1.18

while I was on his train, which wasn't a great idea because I couldn't really leave. We had to sit in silence in the conductor's quarters while I ate my lunch out of an off-brand knockoff lunch box. FIG. 1.19 It couldn't even keep the ceviche cold so I was crying into my lukewarm ceviche.

What I'm trying to say is, I've experienced death. As *mortals*, humans have a set life span. While the average *life expectancy* greatly differs from place to place (from 82.6 years in Japan to 0.0 years in the lava of a volcano), every human is going to die. Sorry, bitches. But don't let it sneak up on you! Here are this spring's most glamorous ways to die!

FIG. 1.19

# THIS SPRING'S MOST GLAMOROUS WAYS TO DIE

Don't die like any old girl! It's
so embarrassing when
you die in the *exact* same way
as another girl. Go out in
style in these fun, flirty ways
to expire!

### 10. COMA & EXTENDED VEGETATIVE STATE

Want to lose up to one hundred pounds with no effort whatsoever? Slip into a coma! When your family finally takes you off life support, you'll be the trimmest you've ever been. Better make your funeral open-casket! All the boys are going to want to see this!

### 9. SMALLPOX

Vintage is *in* this spring. Every girl is going to be vying for that elusive retro virus. Before you die an excruciating death from smallpox, you get the cutest little spots all over your skin. No need to wear a sundress when you have skin this trendy—polka dots are the new stripes! Plus, smallpox fits everyone. It's like the Traveling Pants!

### 8. CHOKE TO DEATH ON A LATTE

This is glamorous a "latte" of times!

### 7. KILL YOURSELF AT YOUR BEST FRENEMY'S WEDDING

Now *this* is a fun one! Everyone has that one girlfriend that's a little more enemy than friend. Maybe she stole your boyfriend in college. Maybe her rib cage sticks out more than yours. Maybe she's black and you hate black people. Wait until she's planned her perfect wedding (and invited you, that passive-aggressive bitch!). Right as she's about to kiss the groom (or bride, that lesbian!), run up and shoot yourself in the head in the middle of the altar. This little prank will keep your true friends in stitches. Especially if any of them were hit with shrapnel!!

### 6. CHILDBIRTH

Ladies, you can have it all, ladies. A career, success, and a family! You don't have to give up anything in your life for a family if you die during birth. Giving birth is a great way to lose five to ten pounds immediately, and dying in childbirth is a great way to lose one hundred to two hundred pounds forever! Plus, how cute will it be when your husband has to take care of your child by himself, figuring out how to juggle mac 'n' cheese with grief counseling with ballet lessons? Mr. Mom much!

### 5. KILL YOURSELF AT YOUR EX-BOYFRIEND'S WEDDING

Girls run the world, and any guy who has mistreated you needs to know that. One surefire way to send your ex a message is to kill yourself at his wedding to whatever new slampiece he's found to degrade. Make sure to wear something form-fitting and sexy to the wedding (lavender sheaths are in this spring)—you want your ex to know what he's missing. *FOREVER.*

### 3. RECTAL CANCER

And who says women can't be funny? It's taken years, but women are finally being recognized as the true comic forces that they are. Show that you're just as funny and crude as any man by dying from butthole cancer. Men, look out—women are the queens of potty humor now! And we *sit down* to do it!

### 2. PROSTATE CANCER

It's the twenty-first century—women can do whatever men can do! Up until now, girls have been led to believe they can only die from "ladylike" cancers like breast and ovarian. If women can live like men, they can die like men. Get prostate cancer—show your husband who wears the chemo smock in your relationship!

### 4. DIE DURING SEX

This is where things get a little NSFW! There are plenty of kinky sex moves you can do to please your man (the Tommy Tutone, the Chicken of the Sea, the Reverse Schiavo, et cetera). But how *kinky* will it be when you die below (or on top of, you Annie Oakley you!) him during sex? There are a few ways to accomplish death during sex: you could choke, drown, get hit by a train, or your guy could stab you in the heart with his extremely pointy dick! Pick the way that's right for you! It's important that you feel comfortable while you're dying during sex.

### 1. CAPITAL PUNISHMENT

And the most glamorous way to die this spring? Capital punishment! Can anyone say "last meal"? Let's just hope it's on one of your carb cheat days!

# Single-Celled Orgasms . . . Excuse Me, That's Organisms! Oops, My Bad! ;)

Oops, what a SEXY TYPO! ;) I call those "sexpos," which is short for "sexpographical error." Feel free to use "sexpographical error" as part of your daily sexicon ("sexually charged lexicon"). Here's a cheat sheet to remember the difference between organisms and orgasms! FIG. 1.20

FIG. 1.20

## ORGANISMS *vs.* ORGASMS

| ORGANISMS: | ORGASMS: |
|---|---|
| are any contiguous living system (such as animal, fungus, microorganism, or plant). | are the sudden discharge of accumulated sexual tension during the sexual response cycle, resulting in rhythmic muscular contractions in the pelvic region and characterized by sexual pleasure. |
| FEMALE ORGANISMS: | FEMALE ORGASMS: |
| exist. | don't exist (MORE ON THIS LATER). |

# You and Me Baby Ain't Nothing but Mammals, Reptiles, Invertebrates, Birds, Fish, or Amphibians or Other Various Animals!

Breaking news: animals are cute. The method of naming and classifying the animal kingdom is called *taxonomic hierarchy*. The hierarchy is as follows, from largest to smallest: *Kingdom, Phylum, Class, Order, Family, Genus,* and *Species*. There are a lot of fun (and flirty!) mnemonic ways to remember the sequence of the taxonomic hierarchy. One of the most popular ones is *King Phillip Came Over From Great Spain*. Now, I don't want to be too gossipy, but let's just say that's not the only thing King Phillip *Came Over!* ;) And here's a fun one I just made up: *Kiss Penis-Cocks On Full Gag Speed.* That's also a good mnemonic to remember how to properly give (pardon my French) fellatio! I really apologize for the crudeness of the mnemonic, but it's the only way to remember. NEVER FORGET! I didn't make that phrase up, I stole it from 9/11. :)

Get back on track, Meg! No one has ever called me Meg, but that's my nickname for myself when I have to give myself a slap on the wrist. "Finish your chapter, Meg!" "Go to spin class,

24

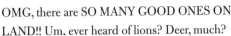

Meg!" "Stop peeing in the bed when you're drunk, Meg!" "You don't know me, Meg, I'm going to pee wherever I want." "Meg, I'm sorry, but you are just being out of line." "That's it, Meg, I'm going to pee on you." And then I wet my pants and go to bed. Animals are an amazing part of the world we live in. My favorite animal is the fly, because its name is what it does.

Animals are so lucky, they get to wear animal prints at all times! Take a look at this little feature, the most fun animal prints.

## Most Fun Animal Prints

Zebra

Tiger

Marissa

There are so many animals that it's going to be hard to cover everything, but I'm going to try to get all the really important kit-kits and fishies in there!

## Land Animals

OMG, there are SO MANY GOOD ONES ON LAND!! Um, ever heard of lions? Deer, much?

## Air Animals

OMG, while I was writing the "land animals" section I TOTALLY FORGOT I had this treat coming up!! And then now this is like a fun surprise! It's like when you leave five bucks in your skinny jeans from last season and then you put them on again and you have the five bucks in your pocket and you also lost a little weight so they look *amazing* and you love *love*!

My favorite air animals are birds!!!

## Swimming Animals

Fishies, whale-ies, sharkypoos! Also, tadpoles! OMG!

# Evolution

The time was late 1831. The place was England. Let me take you back: there was tea in the kettle, the 1832 calendars were coming out and going on sale for next year, slaves were at an all-time high, 1831 calendars were at an all-time low!

A young man named Charles Darwin was sailing around the Galapagos Islands in a boat called the HMS *Beagle* (how cute is THAT, a hot twenty-two-year-old in a boat called the *BEAGLE*!) and he observed that the cute little *finches* (oh my God, this is too cute!!!! brb I have to go JERK OFF) had developed changes that made them better suited to their environments, called *adaptations* (FYI, back! Got a quick jerk sesh in! Thought about fat Val Kilmer the whole time, how weird is that!).

Darwin realized that those birds more suited to the Galapagos environment were more likely to survive and pass their genes and traits on to their offspring. This is called *natural selection* and is a huge part of *biological evolution*. If a species doesn't change, it becomes *extinct*, like the dinosaurs or huge seventies bushes. Like, here's an example. Girls that are shaved clean in their DOWNTOWN DISNEY like a newborn baby with *alopecia* (disease where all your hair falls out—sign me up, please, no more waxing, baby! Do I make you horny, baby!! ;)) are more likely to *mate* and then pass on their genes than girls with big ol' tufts of hair. Then the shaved girls' babies will all be shaved as well because they'll have inherited the trait, and then the species will have *evolved* to all have hairless Beasts of the Southern Wild (vaginas).

This fun magazine chart will help your ladybrain grasp the concept of evolution! FIG. 1.21

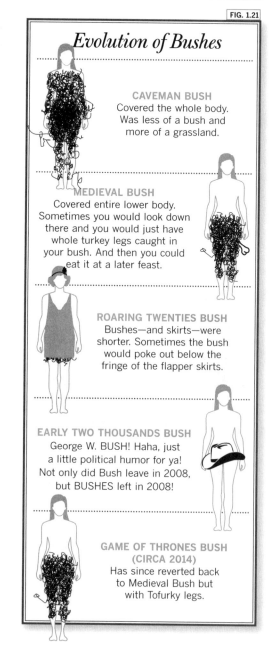

FIG. 1.21

## Evolution of Bushes

**CAVEMAN BUSH**
Covered the whole body. Was less of a bush and more of a grassland.

**MEDIEVAL BUSH**
Covered entire lower body. Sometimes you would look down there and you would just have whole turkey legs caught in your bush. And then you could eat it at a later feast.

**ROARING TWENTIES BUSH**
Bushes—and skirts—were shorter. Sometimes the bush would poke out below the fringe of the flapper skirts.

**EARLY TWO THOUSANDS BUSH**
George W. BUSH! Haha, just a little political humor for ya! Not only did Bush leave in 2008, but BUSHES left in 2008!

**GAME OF THRONES BUSH (CIRCA 2014)**
Has since reverted back to Medieval Bush but with Tofurky legs.

# Evolution vs. God

God. Where to even begin? Literally! Do I begin at the *big bang*, or the day that God made everything? Science or religion? It's been a debate for centuries, decades, even.

As a science expert and premarital-sex enthusiast, I don't have much room for religion in my life. But I totally understand that some ladies do need to think that there's an invisible man watching them at all times. That gets me a little hot, too! You're telling me there's a guy who's *in everything* at all times? Maybe that's that tingle I've been feeling constantly in my "Temple of Doom." ;)

## What RELIGION is RIGHT for your BODY TYPE?

There are so many religions these days. It can be hard to know which is right for you. But it's easy to know which is right for you if you know your body type! Use this quick guide to figure out what faith you should immediately convert to and dedicate the rest of your life to and change your children's religion to if you have children.

**PEAR-SHAPED:**
With full hips and butt, you'll want to draw attention up to your toned shoulders, arms, and hands. Go with Catholicism—it's totally "in" to wear huge crucifixes around your neck, which will highlight your pretty clavicles as well. Plus, what draws more attention to arms and hands than a cute case of stigmata? Get the name-brand kind or make your own!

**APPLE-SHAPED:**
Apples have big bellies and traditionally leaner legs. Try Buddhism, which was started by a famous Apple shape. Because Buddha himself was an Apple, you'll fit right in. And if you need to lose that belly quickly, self-immolate! Buddhists have been using that secret weight-loss technique for ages!

**BOYISH:**
Straight up and down? Make your religion Hinduism. The sari will fit your thin frame well (Gandhi-style hunger strike anyone?), but also a major tenet of Hinduism is that life is inherently a cycle of suffering through death and rebirth, and obviously you would believe that life is suffering if you were born with no boobs!! Maybe in your next life you'll be Kate Upton, sister. :)

**PLUS-SIZED:**
You'll want a religion that covers up all your problem areas. Go with a religion that requires women to cover up as much as possible. This means anything fundamentalist—Islam, Orthodox Judaism, Mormonism . . . take your pick! There are so many good options for plus-sized women these days! We've come a long way!

As I mentioned previously, I myself am a Jew. I can't say I'm proud of the fact. It's tough being a Jew in modern society: I'm extremely dark-complexioned for a white person, and Louis Vuitton doesn't make a good midsize sack for Jew-gold.

Because I'm a Jew, I was raised with all the strong, applicable traditions of an Abrahamic religion. And who's a stronger female role model than Eve! Even Eve had some guy troubles, though. Here at *Science . . . for Her!*, we've got a first look at Eve's dating profile. FIG. 1.22

Speaking of dating sites, let's talk about me for a second for once, God! Just a little update: I think I'm making a little headway with Xander! We've entered into a long-distance relationship: his lawyer recently contacted me and said I was required to stay five hundred yards away from Xander at all times! Everyone knows long distance is tough, but I had no idea it would be this tough! Hmm, maybe I can start calling his lawyer a lot to make Xander jealous? Such a great idea, me!

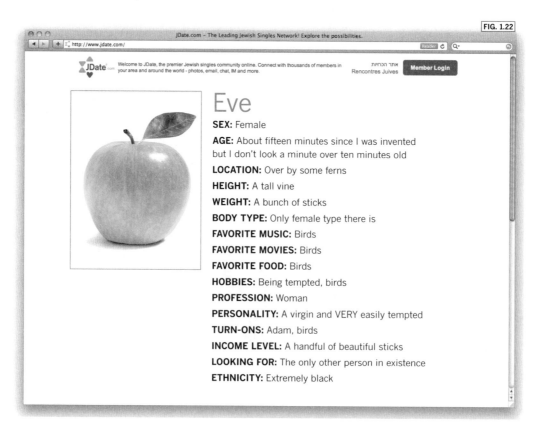

FIG. 1.22

JDate.com – The Leading Jewish Singles Network! Explore the possibilities.

http://www.jdate.com/

JDate.com Welcome to JDate, the premier Jewish singles community online. Connect with thousands of members in your area and around the world - photos, email, chat, IM and more.

אתר הכרחית Rencontres Juives

Member Login

## Eve

**SEX:** Female

**AGE:** About fifteen minutes since I was invented but I don't look a minute over ten minutes old

**LOCATION:** Over by some ferns

**HEIGHT:** A tall vine

**WEIGHT:** A bunch of sticks

**BODY TYPE:** Only female type there is

**FAVORITE MUSIC:** Birds

**FAVORITE MOVIES:** Birds

**FAVORITE FOOD:** Birds

**HOBBIES:** Being tempted, birds

**PROFESSION:** Woman

**PERSONALITY:** A virgin and VERY easily tempted

**TURN-ONS:** Adam, birds

**INCOME LEVEL:** A handful of beautiful sticks

**LOOKING FOR:** The only other person in existence

**ETHNICITY:** Extremely black

# Biology Recap

Oh my GOSH, girls! I can't believe it! We've made it through one whole chapter! I don't know about you, but I am exhausted, famished, and bored as *hellllo KITTY!* This was like a *Cosmo* but so, so much more boring! We deserve a treat!!

## Best Ways to Treat Yourself After You Read a Whole *Boring* Biology Chapter in a Women's Science Textbook

**1.** Rent *Sleepless in Seattle* on Netflix and make yourself a Seattle Slush (gimlet with rain water!). This classic movie and cute themed cocktail are just what you need to relax—that chapter was boring as hell!

**2.** Do a little online shopping and snag a hot monokini from a vintage online store.
Monokinis are such a fun, flirty, retro alternative to bikinis! It's a great way to pump yourself up after such a fucking awful time reading a biology chapter in some fucked-up textbook.

**3.** Try a new hobby, like scrapbooking, scuba diving, or wine tasting!
Yeah, definitely wine tasting, you're going to have to get f@($%*ing wasted to just forget how much you wanted to TRULY DIE while reading this GOD-AWFUL CHAPTER. Oh my dear GOD. I am so drunk right now. God. Xander I miss you so much. I'm going to text him. Fuck my knuckles have wine all over them I can't read the digits to Xander's phone number.

# Recap Questions

**QUESTION 1:** Do female organisms exist?

**QUESTION 2:** What is asymmetrical about Xander's face, and why doesn't it matter to me?

**QUESTION 3:** Why are your bangs so fucking ugly, you two-faced bitch?

**QUESTION 4:** Why do good things happen to bad people/pear shapes?

**QUESTION ?:** What number comes after 4 again? I am so bad at math!

**QUESTION OH YEAH, 5:** Well, I guess this is technically Question 6, since the question before was a question.

**QUESTION 5:** That wasn't a question before, that was an explanation, so *this* is question 5! Why are you so fun? Love you, girls!

# 2 Chemistry

# Introduction

You're back! OMG I've missed you so much, baby girl!!!! I know we promised to keep in touch over the chapter break, I'm honestly *so* sorry I didn't write or call or anything. I just got super busy writing this upcoming chapter, and also I've been platonically friend-dating this new guy, Styx (he can tattoo his own thigh, it's so cool!!), I started a juice fast and I was really tired for the first few days when you only get two juices a day . . . oh my God, I've become one of those girls who makes excuses for not hanging out with her girlfriends. I am like gonna burst out into tears. I love you, babe. I'm so glad we're back. You're my best friend. :) Dinner soon, please! :)

I know what you've all been waiting for—Xander update! Well, today's your lucky day, babes! I think it's actually going super well. I ran into him at a CVS six months ago while I was buying tampons and instead of talking to him, I burst out crying and I threw a bunch of AAA batteries from the battery aisle at him (which they made me buy EVEN THOUGH I never took them out of the package) and it sort of sent me into a spiral and then I didn't write for like six months. So yeah, I think we're really working out some of our shit! I love that we're both adults and can just, you know, communicate.

*Chemistry* comes from the Greek *alchemy*, which, loosely translated, means "chemistry." The study of chemistry is all about matter, and boy, does it *matter*! Chemistry is easy when you realize that everything you taste or touch is a chemical. Like, if you've ever worked with bleach, that's a chemical! And if you haven't, what kind of a woman are you! Clean your family's goddamn clothes! If you've ever cooked or baked, that's chemistry. Painted your fingernails? Chemistry. Painted a dog's fingernails? Chem-mutt's-try! Shouted out a chemistry equation while faking an orgasm in the Bucking Bronco sex position (this chapter's featured sex move!)? Chemistry. But also: a *great Friday night!* ;) Yee-HAW! Ride 'em cow-GIRL! My safe word is "MORE, PLEASE!"

I'm going to try to make chemistry as fun for you as possible. Chemistry can seem very abstract and scary, like your dad turning into a smoke spider. But it's actually very concrete and scary, like your dad turning into someone who would divorce your mom after forty-three years of marriage. **FIG. 2.1** After this chapter, I promise you won't fear chemistry any longer. The only acceptable things for a modern woman (Can I get an "I am woman, hear me roar"?!) to be afraid of: mice, rats, mortality, candles, HPV, bugs, your vagina being ugly, aging, ending up alone, your vagina being too long vertically, whole milk.

Sit back, relax, and pour yourself a Malibu Spice Tini-Meanie (Malibu spiced rum, vodka, and Atkins meal-replacement milkshake—yes, those are chemicals, too! ;)), because it's time to *"Let's Get Chemicals! Chemicals!"* (sung to the tune of "Let's Get Physical" by Olivia Newton-John!).

FIG. 2.1

AUTHOR'S NOTE: I should have saved that pun for the physics chapter. *"Let's Get Physics, Y'all!"* is actually like the most perfect thing I've ever thought of. Well, I'll use it there, too.

# Food & Cooking—Yum Yum!

To ease into the extremely difficult science of chemistry, we'll begin with the chemistry of cooking! Most women inherently grasp the science behind cooking without even thinking about it, but we'll spell it out. Not literally—don't worry, ladies! We'll save spelling for another book!

There are three major types of molecules that make up the food that we eat: *carbohydrates*, *lipids*, and *proteins*.

**CARBOHYDRATES:** Carbohydrates (literally hydrates of carbon) are chemical compounds that act as the primary means of storing or consuming energy. The most common carbohydrate for human consumption is sucrose, or table sugar. If you want to be skinny, DO NOT EAT THEM.

**LIPIDS:** *Lipids* is the scientific name for fats. Fats that are liquid at room temperature are often referred to as oil. Lipids in food include the oils of such grains as corn and soybeans, or animal fats, and are parts of many foods such as milk, cheese, and meat. If you want to be skinny, DO NOT EAT THEM.

**PROTEINS:** In food, proteins are essential for growth and survival and vary depending upon a person's age and physiology (e.g., pregnancy). Proteins in food are commonly found in peanuts, meat, poultry, and seafood. If you want to be skinny, DO NOT EAT THEM.

**FOOD:** If you want to be skinny, DO NOT EAT THIS.

**PAPER MENU:** While not *technically* thought of as a food by most scientists, this is a good option if you don't want the calories that come from "food"! When you go to the restaurant, just ask for the menu!

**AIR:** Also a guilt-free part of a balanced breakfast!

**CIGARETTES:** Totally fine! Like air but better tasting!

**FOOD:** Bad!

**NOT FOOD:** Good!

**WE:** ARE BEST FRIENDS!

# Healthy Cookin'

All girls want to stay trim and healthy! Everyone's favorite celebrity chef, *Paula Deen*, has created a health-food cookbook that is basically foolproof.

# PAULA DEEN'S
# HEALTH-FOOD COOKBOOK

**Everyone knows cooking is just chemistry wrapped in practicality wrapped in fun.
It's the turducken of the sciences! Recently, Paula Deen has admitted that she's
had type 2 diabetes for years. Accordingly, she's putting out a cookbook of healthy
food. Here are some excerpts!**

## FRUIT SALAD

**INGREDIENTS:**
1 1-lb. bag Skittles
3 cups ranch dressing

**DIRECTIONS:**
Mix well. Serve at room temperature.

## PAULA'S BROWN RICE

**INGREDIENTS:**
1 pilaf white rice
1 bowl melted Junior Mints

**DIRECTIONS:**
Cover rice in chocolate. Serve with maple
syrup to taste. To splurge, top with a
sprinkle of sausage calzones.

## SCRAMBLED EGG WHITES

**INGREDIENTS:**
1 dozen (12) Cadbury Creme Eggs
2 lbs. Frito crumbs
1 package extra-fat pork lard `FIG. 2.2`
1 pilaf Paula's brown rice

**DIRECTIONS:**
Break the Cadbury eggs and harvest the
crème-filled white centers. Dip them in the
Frito crumbs. Put the lard (make SURE to get
the extra-fat kind or it will be BLAND) in a
frying pan on high heat, and fry the crème
centers until golden brown. Serve on a bed of
Paula's brown rice.

## PAULA'S GARDEN BURGER

**INGREDIENTS:**
3 bags Olive Garden® Endless Breadsticks
12 Olive Garden® Stuffed Mushrooms
1 plate Olive Garden® New! Baked Pasta Romana
with Chicken
4 Olive Garden® Black Tie Mousse Cakes
1 slice American cheese (optional)

**DIRECTIONS:**
Smash all of the Olive Garden® foods together
until they resemble a large patty and top with
cheese. For lower calories, hold the cheese.

## PAULA'S GUILT-FREE
## FAT-FREE® SMOOTHIE

**INGREDIENTS:**
34 lbs. sugar

**DIRECTIONS:**
Put sugar in smoothie glass and drink with straw,
serve chilled in white-wine tumblers, or, for
special occasions, lap from trough. This delicacy
is guilt-free since you can make a conscious
choice not to feel guilty about anything you put in
your body, like Paula does!

`FIG. 2.2`

## BUFFET AND A BURGER

**INGREDIENTS:**

1 burger
1 Las Vegas buffet
Christmas-themed elastic pants (optional)

**DIRECTIONS:**

Go to Las Vegas buffet. Make sure the buffet has burgers, or provide your own. Do NOT walk around the buffet. Get a motorized scooter, or stay in one spot and use a jaws of life to pick some of each buffet food out of the tubs and put it on your burger. Elastic pants are nice because your gupa (gunt-fupa) stays nicely inside the stretchy pants except for a few folds of fat with stretch marks that seep out of the pants.

## PAULA'S GUILT-FREE® PEANUT BUTTER AND JELLIES

**INGREDIENTS:**

1 peanut
18 sticks butter, mashed
1 pair Jellies shoes

**DIRECTIONS:**

Cover the shoes with butter and top with the peanut, then eat the shoes. If you eat shoes it's like you're exercising so it's VERY healthy.

## PAULA'S GUILT-FREE® PIZZA PANTS

**INGREDIENTS:**

10' x 20' swath of pizza
More pizzas to use as pepperonis on the pizza
Stuffed mushrooms
FYI the mushrooms are stuffed with smaller pizzas
Smuckers Magic Shell ice cream topping
Rolos
Coca-Cola
3 bags gummy bears
Fondue
Caesar salad dressing
Wood chips (as a thickener)
Grenadine syrup
Butter-flour mixture
Pizza Pockets
1 sewing machine
1 sewing pattern for pants (size XXXL)

**DIRECTIONS:**

Mushrooms are a vegetable and there are definitely some mushrooms on that pizza so technically they are HEALTHY-style pizza pants. Take the really big pizza. Put all of the other ingredients on the pizza. Pour the Coke on the pizza. Dip the pizza in the fondue, and resist eating it before you make it into pants, no cheating!!! Sew that pizza into pants using the machine and the pattern. Make sure to sew in some pockets so you can keep a few extra Pizza Pockets in your pizza pockets!!!! Then eat your pants!!!!!!!!!!!

## PAULA'S GUILT-FREE® TURTURTURDUCKDUCKENDUCKEN

**INGREDIENTS:**

3 turduckens FIG. 2.3

**DIRECTIONS:**

Stuff a turducken in a turducken in a turducken. While you're waiting for it to cook, make your fat niece make you some pizza pants while you're watching *Pawn Stars* and eat your pants and then slap your niece.

FIG. 2.3

## INSULIN AU GRATIN

**INGREDIENTS:**

1 insulin shot
1 15-lb. block of cheddar cheese

**DIRECTIONS:**

Bury insulin shot in cheese. When you're going into a diabetic coma, just eat your way to the shot!! Eat the cheese fast or you'll die!!!!!!!!!!!!!!!!

## SPARKLING WATER

**INGREDIENTS:**

1 glass sparkling water
1 ham

**DIRECTIONS:**

Put ham in water.

# A Lady on the Streets, but a Scientist Also on the Streets!

We at *Science . . . for Her!* pulled aside a bunch of ladies on the streets to see how they bring chemistry into their day-to-day lives!

*Science . . . for Her!*'s

## LADIES ON THE STREETS

"Who are you?"
—Tina, 25

"I can't talk right now,
I'm late for work."
—Michelle, 24

"Why are you touching me
and calling me Xander?
Get the fuck off or I'm gonna
call the cops."
—Denise, 19

"You smell like rotten
daiquiris and you have vomit
on your scarf."
—Elizabeth, 34

"Ma'am, why are you
harassing these people? We're
going to have to take you
downtown."
—Officer T. Malcolm, LAPD,
50s-ish

"If I suck your wiener off will
you let me go even though
I am so wasted on expired
daiquiri mix that I stole from
the food bank when I was
volunteering there???"
—Megan Amram, 26

"Honestly? Yes." —Officer T.
Malcolm, LAPD, 50s-ish

# Household Chemicals

Cleaning is a type of chemistry! Cleaning chemistry happens all around us and we don't even think about it. I am so crazy about cleaning stuff. I am so crazy that sometimes it cuts into my daily routine. I was going to see a doctor about it but I couldn't leave my house because the doorknobs were dirty and possessed by the devil unless I touched them 666 times and whinnied like a pony, so I'm pretty sure it just cleared itself up.

I often find that I buy too much bleach, Windex, and other common chemicals. Instead of letting them just sit around the house gathering dust (and then you have to clean the cleaning supplies and that's ironic!), I've created some fun cocktails you can make that help use up those pesky chemicals.

## HOUSEHOLD-CHEMICAL COCKTAILS

Boring old cocktails are so both early 2004 and late 2009! Spice up your daily routine by using commonly found chemicals in your house to add a little pizzazz to your libations!

WARNING: IF YOU DRINK THESE, YOU WILL DIE.
(see LIFE CYCLE section)

**"SEX ON THE BLEACH!"**

**Ingredients:**
vodka, peach schnapps, orange juice, cranberry juice, bleach.

**Calories:** unknown.

*WARNING: IF YOU DRINK THIS, YOU WILL DIE.*

**"ANTIFREZZY NAVEL"**

**Ingredients:**
peach schnapps, orange juice, antifreeze.

**Calories:** unknown.

*WARNING: IF YOU DRINK THIS, YOU WILL DIE.*

**"GIN AND DRANO"**

**Ingredients:**
gin, Drano.

**Calories:** unknown.

*WARNING: IF YOU DRINK THIS, YOU WILL DIE.*

# Acids & Bases & Tigers & Lions, Oh My!

Acids and bases are why some foods taste sour (acids, like lemons) and bitter (bases, like semen—even the sweetest-tasting male ejaculate tastes like undercooked escarole to me! But I have still swallowed twelve times and counting!). Calling something an acid or a base refers to its pH, which measures the hydrogen concentration in its solution. Pure water has a pH very close to 7 at room temperature. I had a boyfriend who was a 7 at room temperature! FIG. 2.4 In the cold he was only a 3. :(

If you walk around your kitchen (and you should be, at all times, in heels and nothing else!), you will see many household acids and bases that are VERY important to cooking/cleaning/being a woman.

**IMPORTANT ACIDS:** vinegar, molasses, coffee, lemony semen (I tried dabbing a little lemon juice onto the tip of my boyfriend's doodle so that when he django'd in my mouth it would taste HALFWAY DECENT, but it 1) didn't help much with taste and 2) made him scream in pain), FIG. 2.5 Lemony Snicket (that's the same as lemony semen, it's just a cooler name, you put some LEMON on the tip of his SNICKET)

FIG. 2.5

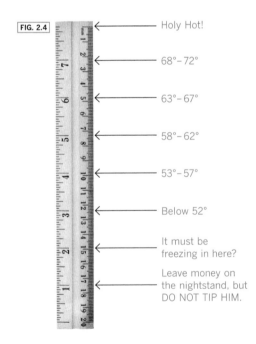

FIG. 2.4

← — Holy Hot!

← — 68°–72°

← — 63°–67°

← — 58°–62°

← — 53°–57°

← — Below 52°

← — It must be freezing in here?

← — Leave money on the nightstand, but DO NOT TIP HIM.

**IMPORTANT BASES:** baking soda, Ivory soap, bleach, Tide, dishwasher detergent, normal semen—BLECH!, egg-yolk semen (same as Lemony Snicket but with egg yolks, super disgusting, when I did this I threw up on his dick!), vomit dick

# Alcohols

In chemistry, an alcohol is an organic compound in which the hydroxyl functional group (-OH) is bound to a carbon atom. In life, alcohol is the best most fun thing and the only thing that makes living fun! FIG. 2.6 In fact, I'm already like two shots and a bottle of wine deep. And maybe this is just the booze talking, but hello, I am a bottle of booze and some shots of booze!

There are differences between alcohol as defined by chemistry and the kind of alcohol that gets you pregnant. The type of alcohol in alcoholic drinks is *ethanol*, the chemical compound $CH_3CH_2OH$. A good mnemonic device to remember this compound is "CHA-CHACHACHACHACHAOH!" Like a flamenco dancer doing a big dance and then realizing that she forgot to wear her turban with fake fruit on it.

Wine is a good alcohol! Wine is made from old grapes. When a grape sits for years and years and years, it becomes alcoholic, just like some grandpas!! I am so clumsy. Sometimes I spill red wine down my throat/mouth like six times a day every day. Whoops!

FIG. 2.6

## BAD ADJECTIVES TO USE AT A WINE TASTING

Curdled

Like white wine, but red

Boiling

The N-word

Like booze-gravy

Wife material

Tastes like what I had as a kid in the orphanage

Deep-dish

Chubby

Shrimp-flavored

Circumcised

Like a grape lover's semen

Like bad wine

Winey

Gay

# The Science of *Chemistry* ;)

You've probably been distracted this whole chapter by the fact that "chemistry" is SUPPOSED to mean the tingles you feel in your lady-lumps FIG. 2.7 and gal-valleys when you're near your crush. Well, that's what we're going to talk about now! That's *chemistry*-chemistry!

The smell of other humans is a really big part of chemistry-chemistry. Smell is honestly one of my top five favorite senses. It's like right up there with taste, feel, sight, and sound. Actually, you know what? *Fuck sound.* It knows what it did. Actually, fuck sight, taste, and smell too. I'm rolling with just feel. That way I'll be able to enjoy fuzzy puppies and not have to smell their poop or look at their stupid faces or hear their dumb barks. Speak English, idiots! FIG. 2.8

Humans give off something called *phero-mones*. They are chemicals that trigger responses in members of the same species. Certain phero-mones, called *axillary steroids*, are produced by

FIG. 2.7

FIG. 2.8

the testes, ovaries, apocrine glands, and adrenal glands. That being said, love isn't all just sniffs 'n' smells. Love is extremely tough, even in the best circumstances. You gals have been on the whole Xander journey with me. You get it. One day you're deeply in love, and the next day you're trying to pelt him in the eyeballs with unopened batteries because if he doesn't want to look at

you naked he shouldn't be able to look at any woman naked. To prove how much love SUCKS sometimes, I've ripped this right out of the *New York Times*, which I got from a Starbucks! (I get like *everything* from Starbucks. At Starbucks, it's so funny that the coffee is five dollars while the napkin dispensers are free!)

# The New York Times

## Corrections

The *New York Times* would like to issue corrections for the wedding announcement of Mr. Adam Penview to Ms. Katie Jasper that ran in yesterday's paper.

We incorrectly identified in the announcement that Mr. Adam Penview and Ms. Katie Jasper were married at the Church of the Holy Trinity in Manhattan. They were married at St. Brigid's.

Additionally, we inaccurately wrote that the groom "attended Cornell University and double-majored in English and Humping Other English Majors' Girlfriends." Mr. Penview, in fact, majored only in English. He did not major in "Humping Other English Majors' Girlfriends," as that is not currently an existing track of study at Cornell or any other accredited university.

We erroneously wrote yesterday that Mr. Penview was the "son of Dr. Ryan Penview, a third-generation ophthalmologist, and Mrs. Claire Penview, a Zuckerberg-ass beaver-bitch." Mrs. Penview practiced law in New York State until 2004, and is considered by many to be a friendly and beautiful member of her community, bearing no resemblance whatsoever to Mark Zuckerberg or his rear end. "Beaver-bitch" is not a profession.

We mischaracterized the bride as having worn "a peace [sic] of shit mayonnaise tent. Also,

you know how sometimes people see the Virgin Mary in stuff? It was like that, except you could see Hitler in the wedding dress, but specifically because she had hand-embroidered a picture of Hitler in her dress." In truth, Ms. Jasper wore Amsale.

*Mr. Penview clearly standing on the campus of Cornell University in 1998.*
*Image © some English major's girlfriend probably*

We incorrectly noted yesterday that the couple met "while the bride was dating the totally devoted and now very successful Assistant *New York Times* Wedding Section Editor Dan Gould. Dan guesses he just wasn't good enough for you Katie, because you had to go hump that piece of shit Adam Penview that you met at the English major mixer while Dan left Ithaca for the weekend to go to his Nana's 90th birthday in Needham, Mass. Dan even brought you back a hat from the birthday party that said 'Ethel's Doin' It for One Night Only.' FYI, Adam, it counts as incest if you sleep with another English major's girlfriend, since English majors are BASICALLY BROTHERS. Nana will live forever!!!" Though factually correct, the *New York Times* apologizes for the way in which Mr. Penview and Ms. Jasper's first meeting and relationship history were portrayed. Additionally, the hat actually read "Ethel's Not 90 . . . She's 89.95 Plus Tax." She has since passed.

We wrote that the couple was married by "a dildo with googly eyes, which is the kind of freaky stuff Adam is into." They were married by Father Norman Murray. Additionally, Mr. Penview is into regular stuff.

Yesterday, we printed that "God Katie sorry I'm writing all this I've taken a lot of Robitussin you are so beautiful. You look like a young Nancy Kerrigan. Oh God I love you." In fact, Mr. Gould had only taken a moderate amount of Robitussin.

We wrote yesterday that "Katie sorry you can't fucking deal with my Jew strength you blimp-bitch." Ms. Jasper, in truth, can deal with said Jew strength.

Yesterday, we wrote, "Katie, I'm so sorry, oh God, please come back, I think you can get weddings annulled really easily, it's like a five-second-rule-type thing. Also Adam, I'm really sorry about all the googly eye stuff, I actually think you're a pretty good guy, I once had a dream that you were a creature that had your torso and face but a gay man's body and you saved me from Gwyneth Paltrow in 'Contagion.' Granted, I had taken a lot of Robitussin that night, but when I woke up I did feel sincerely indebted to the top half of you." We don't know where to start. The *New York Times* is just so sorry for this entire paragraph.

Hey just kidding about all these things that we said we retracted, we're just the dumb ol' *New York Times* what do we know about anything but hard candy and old wrinkly balls!!!!!

We wrote in the paragraph above that "Hey just kidding about all these things that we said we retracted, we're just the dumb ol' *New York Times* what do we know about anything but hard candy and old wrinkly balls!!!!!" In actuality, Dan Gould just broke back into the Corrections department and stole the computer that I'm writing this on. He has been promptly escorted from the building and arrested. In addition, the *New York Times* is an extremely current print newspaper that offers breaking unbiased news and fun crossword puzzles. Fuck Dan Gould.


We retract "Fuck Dan Gould."


We apologize for these mistakes.

# Sexiest Molecules

Some molecules look like dicks! Here are my picks for this winter's sexiest molecules!

## WINTER'S *Sexiest* MOLECULES

**4-Vinylguaiacol**
Also known by his stage name 2-methoxy-4-vinylphenol, this fella is the aroma component of buckwheat. The aroma can remind people of apples, spices, peanuts, wine, cloves, or curry. But also, the $H_3CO$ on the end looks kind of like a dick!

**Acetonitrile**
There's no mistaking it with this guy—that is a dick! That little carbon-nitrogen bond at the end is unmistakably a dick. And acetonitrile is colorless, volatile, flammable, and toxic. Just like a dick. ;)

**Boron trifluoride**
Depending on how you look at this, it either looks like a dick, or two dicks. So make sure to look at it the way that it's two dicks! Because two dicks are better than one (dick)!

**Water**
Even water looks like dicks on a molecular level! That's why, when I'm thirsty, I ask for a glass full of millions of microscopic dicks!!

# Who Wore It Best?

There's nothing more embarrassing than showing up wearing the same atom as another molecule! Some molecules look almost indistinguishably similar. But one is always knocking it out of the park! Gals—which molecules wore it best?

## *Who Wore It Best?*

### Sodium Chloride vs. Barium Chloride

Oops! Looks like NaCl (table salt) and BaCl (a caustic powder used in labs) both showed up to the party wearing the same chloride atom! Sodium keeps chloride a little more modest, though. Chloride should be treated as an accessory, not a statement.

| NaCl | BaCl |
|------|------|
| 56% | 44% |

### Ammonia vs. Water

Ammonia, used in many of your favorite cleaning products, is wearing hydrogen just like water. Sorry, ammonia, but water blows you out of the . . . oh, you know. ;) Water is putting the gorgeous hydrogen on display, while ammonia is burying it by going a little overboard.

| Ammonia | Water |
|---------|-------|
| 38% | 62% |

### Formaldehyde vs. Diamonds

This is a tough one. Both formaldehyde (the chemical used to keep living flesh fresh after death) and diamonds (Ooooohhh!! Sparkly!!) are looking pretty trim on the red carpet with their carbon atoms. Diamond is looking a little bloated, however. Make sure your carbon atoms *fit* before you walk out the door. No one's gonna care about your carbon if it makes you look fat. Formaldehyde wore it best! Diamonds, you look like a fat pig!

| Formaldehyde | Diamonds |
|--------------|----------|
| 87% | 23% |

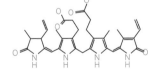 

### Poop vs. Kim Kardashian

Let's just say: both these are messes!! Both are wearing *bilirubin*, which gives them their brown coloration. Kim is just trying too hard, as always. Poop wore it best!!

| Poop | Kim |
|------|-----|
| 81% | 19% |

# The Periodic Table of the Elements

The periodic table organizes chemical elements based on their atomic numbers, electron configurations, and recurring chemical properties. Before the periodic table was organized by Dmitri Mendeleev in 1869, it was just thrown together all willy-nilly! You couldn't find hydrogen if your life depended on it, and nitrogen was stashed under the sofa in the living room. **FIG. 2.9** Everything was all over the place, like how druggies hide drugs in all the little nooks and crannies of their houses. Like, for example, where do you hide your drugs? Please write or e-mail me the answer and don't skimp on the specifics!

A chemical element is a pure chemical substance consisting of one type of atom distinguished by its atomic number, which is the number of protons in its nucleus. Elements are presented in order of increasing atomic number (number of protons). As of this writing, which is happening at 1:55 P.M. on May 26, 118 elements have been discovered or artificially created in a lab. I'm in a pizza parlor. I just ordered a kale salad and, to be honest, Xander is still very much on my mind. I told the pizza hostess that I needed a table for two because my boyfriend was going to meet me. I don't know, I guess that's lame. Call me crazy, but I just thought that MAYBE for ONCE Xander would get my extrasensory brain waves that I've been telegraphing to him through the smooth stone etched with the word "BREATHE" I have in my pocket and figure out that he had to be at dinner with me even though we broke up seven months ago and I shaved my name into his golden Lab. I know—that's what every ex-girlfriend thinks. :(

**FIG. 2.9**

7 N

# *Periodic*
# TABLE SETTINGS

**The periodic table is one of the most important building blocks of chemistry. But boy does it look drab! Here are five fun lady-ways to spice up a boring periodic table.**

**1**    Arrange the boxes so the letters spell something cute like "Fe Rn." Ferns are so cute!

**2**    Make all the boxes into a flower! Doesn't matter which box goes where, as long as it's a pretty flower!

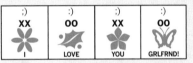

**3**    Rearrange the table into one very long line. This will be slimming no matter what you're wearing!

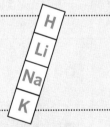

**4**    Erase all the letters in the boxes and make them all have little Hello Kitty heads in them instead.

**5**    Turn the periodic table into one big Sudoku and then have your husband do it for you!

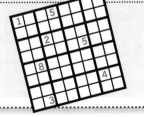

# THE PERIOD! ICK! TABLE

**What would a women's science book be without a spotlight on periods? In this segment, I show a periodic table where every element represents a different aspect of menstruation!**

| | | | | | | | | |
|---|---|---|---|---|---|---|---|---|
| **1** H<br>Hormonal | | | | | | | | |
| **3** Li<br>Light flow | **4** Be<br>Behavioral changes | | | | | | | |
| **11** Na<br>Natural odor | **12** Mg<br>15 **mg** of iron per menstruating day (recommended) | | | | | | | |

| **19** K<br>Kotex | **20** Ca<br>Caked-on blood on a pad | **21** Sc<br>Scabs, It looks kind of like wet | **22** Ti<br>Time of the month | **23** V<br>Vagina | **24** Cr<br>Cramps | **25** Mn<br>"Mnstrtn" is what it would be spelled like w/out vowels | **26** Fe<br>Feeling gross | **27** Co<br>Cold or hot flashes |
|---|---|---|---|---|---|---|---|---|
| **37** Rb<br>RB's (Arby's) craving | **38** Sr<br>Srsly this is gross | **39** Y<br>Yucky | **40** Zr<br>Schnoz redness (acne) | **41** Nb<br>NBD, it's just a period | **42** Mo<br>Mood swings | **43** Tc<br>Bitchy | **44** Ru<br>Ruins your pants | **45** Rh<br>Rhesus monkeys also have them |
| **55** Cs<br>CSI, More blood than an episode of | **56** Ba<br>Bad mood | **57-71** | **72** Hf<br>Bushfire, Red as a | **73** Ta<br>Tampon | **74** W<br>Wide-set vagina | **75** Re<br>Really wide-set vagina | **76** Os<br>Oscillating moods | **77** Ir<br>Irritability |
| **87** Fr<br>Frothy discharge | **88** Ra<br>Raging hormones | **89-103** | **104** Rf<br>Jarfuls of uterus-jam | **105** Db<br>Bloodbath | **106** Sg<br>It's gross | **107** Bh<br>Turn into a dumbhead | **108** Hs<br>Smears on outer thighs | **109** Mt<br>MT (empty) vagina |

| **57** La<br>Labial sensitivity | **58** Ce<br>Cervical mucus | **59** Pr<br>Prone to crying | **60** Nd<br>Ends in menopause | **61** Pm<br>PMS | **62** Sm<br>Smudges on your inner thighs | **63** Eu<br>Euphoria and/or depression |
|---|---|---|---|---|---|---|
| **89** Ac<br>Acting super crazy | **90** Th<br>Thinning cervix | **91** Pa<br>Panty liners | **92** U<br>Underwear skid marks | **93** Np<br>NPR has done radio pieces on it | **94** Pu<br>P.U., this pad isn't scented | **95** Am<br>Amram, gets one, Even I, Megan |

Xe

| | | | | | | | | | | | | | | | | | 2<br>**He**<br>**He**avy flow |
|---|---|---|---|---|---|---|---|---|---|---|---|---|---|---|---|---|---|

| 5<br>**B**<br>**B**lood | 6<br>**C**<br>**C**rimson wave, Riding the | 7<br>**N**<br>**N**ot pregnant | 8<br>**O**<br>**O**vulation | 9<br>**F**<br>**F**our-week cycle | 10<br>**Ne**<br>**Ne**w uterine lining |
|---|---|---|---|---|---|

| 13<br>**Al**<br>**Al**so known as menses | 14<br>**Si**<br>**Si**milar in appearance to miscarriage | 15<br>**P**<br>**P**ad | 16<br>**S**<br>**S**tained sheets | 17<br>**Cl**<br>**Cl**ean, Not | 18<br>**Ar**<br>**Ar**eola tenderness |
|---|---|---|---|---|---|

| 28<br>**Ni**<br>**Ni**ght sweats | 29<br>**Cu**<br>**Cu**nt" is a rude thing to call it, "Bloody | 30<br>**Zn**<br>Qui**zn**os cravings | 31<br>**Ga**<br>**Ga**s and bloating | 32<br>**Ge**<br>**Ge**t it for about three to five days | 33<br>**As**<br>**As** long as seven days, though | 34<br>**Se**<br>**Se**e your doctor if it lasts longer than seven days | 35<br>**Br**<br>**Br**ownish-red tissue | 36<br>**Kr**<br>**Kr**eplach, Similar consistency as |
|---|---|---|---|---|---|---|---|---|

| 46<br>**Pd**<br>**P**D**Q**, get me a towel | 47<br>**Ag**<br>**Ag**itated bowels | 48<br>**Cd**<br>**CD**s of sad singers like Jewel on loop | 49<br>**In**<br>**In**convenient | 50<br>**Sn**<br>**S**anitary **n**apkin | 51<br>**Sb**<br>**S**uper **b**itchy | 52<br>**Te**<br>**Te**n ml of menstrual fluid/day | 53<br>**I**<br>**I**ntestinal distress | 54<br>**Xe**<br>Se**x**? **E**w! Messy! |
|---|---|---|---|---|---|---|---|---|

| 78<br>**Pt**<br>**Pt**ooey, there's blood coming out of my box! | 79<br>**Au**<br>**Au**nt Flo | 80<br>**Hg**<br>**H**. **G**. Wells couldn't invent something this crazy | 81<br>**Tl**<br>**TL**C, Need a little extra | 82<br>**Pb**<br>**PB**&J cravings | 83<br>**Bi**<br>**Bi**g ol' pizza cravings | 84<br>**Po**<br>**Po**cket cravings, Hot | 85<br>**At**<br>**At**rocities in your knickers | 86<br>**Rn**<br>U**rn**born egg-baby |
|---|---|---|---|---|---|---|---|---|

| 110<br>**Ds**<br>Bloo**ds**oup | 111<br>**Rg**<br>U**rg**e to kill someone | 112<br>**Cn**<br>A**cn**e | 113<br>**Uut**<br>Geett**t** oo**uu**t**t** of my uterus, lining! | 114<br>**Fl**<br>**Fl**ooding my undies | 115<br>**Uup**<br>**Uup**s, I bled on your couch | 116<br>**Lv**<br>Vu**lv**a-weeping | 117<br>**Uus**<br>"G**uuu**s**s**sh" is the sound the blood makes | 118<br>**Uuo**<br>"Sq**uuo**oosh" is the sound of me sitting down |
|---|---|---|---|---|---|---|---|---|

| 64<br>**Gd**<br>**G**-**d** created it to punish Eve in the Bible | 65<br>**Tb**<br>**TB** for your vagina (vagina coughing up blood) | 66<br>**Dy**<br>**Dy**ing eggs | 67<br>**Ho**<br>**Ho**le leakage | 68<br>**Er**<br>**Er**ratic emotions | 69<br>**Tm**<br>Mido**l**™ | 70<br>**Yb**<br>Cr**yb**aby | 71<br>**Lu**<br>**Lu**mpy panty-custard |
|---|---|---|---|---|---|---|---|

| 96<br>**Cm**<br>**C'm**on, don't get toxic shock syndrome | 97<br>**Bk**<br>**BK** (Burger King) cravings | 98<br>**Cf**<br>Pubi**c f**estival | 99<br>**Es**<br>**Es**trogen | 100<br>**Fm**<br>**FM**L, I have my period! | 101<br>**Md**<br>**MD** (Maryland) is a place with periods | 102<br>**No**<br>**No**t for boys | 103<br>**Lr**<br>A**l**l **r**ight if you take Advil |
|---|---|---|---|---|---|---|---|

# Carb-On Feel the Noise!

Carb-on feel the noise! Girls rock your boys! We'll get wild, wild, wild! Wild, wild, wild! (I was just informed by my lawyer that those happen to be the lyrics of a song performed by Quiet Riot. Anyway, whoever "wrote" it, any coincidental similarity to my own prose is purely unintentional in perpetuity.)

Carbon forms the key component for all known naturally occurring life on earth. Carbon is abundant on earth. It is also lightweight and the atom is relatively small in size, making it easier for enzymes to manipulate carbon molecules. Lightweight and small? SOUNDS LIKE MY ASS! I *WISH*! FIG. 2.10

After death, we all decompose back into our carbon components. FIG. 2.11

Science has always relied on innovative men (and women! Can I get a "*What Women Want* starring Mel Gibson!!"), but even our most popular heroes age and die. That's just how the carbon cycle works. Organisms are built-in death machines. All organisms experience biological aging, or *senescence*, if they survive all the other downfalls of day-to-day life (like accidents, illness, and killing yourself because you showed up to a party in the same romper as Alexis).

FIG. 2.10

FIG. 2.11

# CELEBRITIES *are* JUST LIKE US!

**MARILYN MONROE . . . DIES!**
Just like us "normies," celebs like Marilyn Monroe eventually die and decompose into carbon and other life-supporting elements!

**MICHAEL JACKSON . . . DIES!**
So good to see that a huge star like Jacko isn't above dying and being interred in the earth! More like he's "below" it—six feet below it!

**WHITNEY HOUSTON . . . DIES!**
This "I Wanna Dance with Somebody" singer should have been singing "I Wanna Die in Some Tubby!" Because she died in a bathtub!

# The Scientific Method

If you're looking for a fun way to practice your chemistry, you've come on the right face! Use that one with boys! :)

The only way to really grasp chemistry is to get hands-on. You're not going to learn it just by reading about it or drinking a bottle of wine and kissing your gymnastics coach in the locker room. The *scientific method* is a process by which scientists try to test hypotheses. I've adapted some classic experiments to better fit a woman's lifestyle.

# *Classic Science Experiments . . . for Her!*

If you need to make chemistry more fun than I've already made it (just add Cosmos! ;) No but really), here are some hands-on ways to test chemical reactions. I've taken classic grade-school science experiments and offered slightly updated versions for the modern woman.

### MENTOS-AND-DIET-SODA FOUNTAIN

**CLASSIC VERSION**

Put some Mentos in a bottle of diet soda and watch it explode! The numerous small pores on the candy's surface catalyze the release of carbon dioxide ($CO_2$) gas from the soda, resulting in the rapid expulsion of copious quantities of foam.

**FOR HER!**

Only eat Mentos and diet soda for two months and see how amazing you look and feel! Mentos have only ten calories per piece, and diet soda has no calories, so . . . on second thought, only drink the soda. Mentos are heavy, anyway!

### BAKING-SODA-AND-VINEGAR VOLCANO

**CLASSIC VERSION**

This experiment looks crazy but is safe as can be! Find a container that looks like a volcano (or make one out of papier-mâché) and fill it with a tiny bit of baking soda. Pour vinegar in after and watch it explode! The baking soda (sodium bicarbonate) is a base while the vinegar (acetic acid) is an acid. When they react they break apart into water and carbon dioxide, which creates all the fizzing as it escapes the solution.

**FOR HER!**

Take the baking soda and vinegar and also 2 tablespoons unsweetened cocoa powder, 2 ounces red food coloring, 1 cup buttermilk, 1 cup rum, 1 teaspoon salt, 1 teaspoon vanilla extract, ½ cup shortening, 2 more cups rum, 1½ cups white sugar, 1 tall glass rum with ice, 2 eggs, 2½ cups all-purpose flour (sifted), 1 cup milk, 1 rum daiquiri, 5 tablespoons all-purpose flour, 1 cup white sugar, 1 cup butter, and 1 teaspoon vanilla extract and make a red velvet cake! Then take that cake and give it to a homeless person who you want to turn fat and drink your dinner of diet soda! Doesn't that feel amazing?!

### DISSOLVING A BONE

**CLASSIC VERSION**

dissolv ea goddam bone i don't fucking care i drank all the rum that i bought for the red velvet cake. You know what??? you don't even put rum in red velvet cake, i put that in the recipe so i could sip while i was writingthis but instead i drank the whol ething cuz im fuckin over it BORNING BORGNIN SCIENCE an im over u zander

**FOR HER!**

i threw up on my bone fuc kme

# Science . . . for Her . . . for Lesbos!

Speaking of EXPERIMENTING, I'd like to take this opportunity to talk *lesbos*. The term *experimenting* has classically meant *the period in college that a girl thinks she might be a lesbian so she starts dating a woman but doesn't want to go down on her so they break up I always loved you I miss you so much Marissa.* FIG. 2.12 Scientists stole the word *experimenting* from lesbians, just like they stole long coats and thick gloves.

Some people say that being gay is something you're born knowing. Not true, chickadees! Being a lesbian is a choice. Just imagine that being a lesbian is like a Choose Your Own Adventure where the end is always scissoring! :)

FIG. 2.12

# QUIZ:
## *Why Did You Decide to Become a Lesbian?*

**If you're a lesbian, it might be a bit confusing to you as to why you chose to be the way you are! Let this fun, flirty quiz help you discover why you chose to be the lesbian you are!**

1. **As a child, were you molested?**
   **A.** Yes! Bad Daddy!!!
   **B.** Yes! By between one and seventeen big scary men on the street or at parties!!!
   **C.** Yes! Figuratively, by the church!!!

2. **Who do you hate the most?**
   **A.** My parents. They both molested me!!!
   **B.** All men. They're boorish and hairy! P.U.!!!!
   **C.** God. What a sillybilly!!!!

3. **You're at school and you see the girl you have a crush on. You like her because:**
   **A.** She looks like my dad a little bit but isn't as dangerous and she never molested me as a child I love you Daddy I hate you Daddy!!!
   **B.** She's a girl and therefore can't hurt me the way a man could she is weak and soft!!!
   **C.** She's an abomination of God, just like me!!!!

---

*Mostly A's*
You lucky lesbo Lizzy: you were molested by your parents!!! ;) Of course you're going to choose to be a lesbian if your home life was wack-a-doo growing up! Being a lesbo is your way of saying, "I'm a lesbian now— F you, Daddy!"

*Mostly B's*
You hate men!!! Big surprise: men are big ol' stinky Petes!! Men are dangerous, smelly cavemen. Maybe you just want a little sensitive lady-time! You don't like the way men treat you and some men only want to have sex instead of just trying to get to know the person! You chose this path because men have treated you poorly and being a lesbo was a way easier option.

*Mostly C's*
On Sundays you're more likely to be found playing softball than attending church because: you hate God!! Being a lesbian is a great way to say, "F you, G-d!" Sure, he may or may not have created your whole family, but you still aren't too excited about doing everything He says. F you, Sky Daddy!

# Bondage!!!! ;)

Molecular bonding is way boring but not when you think of it as bonding with your fave five girls! *Covalent bonds*: those are your besties; but watch out for the frenemies of the molecular world, the *electrons* that pull apart the bonds! They might invite you to their baby showers, but they are NOT your real friends, they will put gum in your baby-shower-themed mocktails and constantly remind you that Xander will never give you kids because he had a voluntary vasectomy in ninth grade! F you, bitches, those can be reversed!

Atoms, the smallest particles of elements, are formed around a dense central core (called the *nucleus*) with negatively charged pieces (*electrons*) flying around the center. The electrons of different atoms interact with each other and bond. The electrons are negatively charged because they've been thinking about their TURNOFFS! My turn-offs include: being dirty, clinginess, huge felony record, not enough felonies. Exactly one felony is what I'm looking for, boys. But a big one. Make it count.

## *You and Me Baby Ain't Nothin' but Mammals, So Let's Do It Like They Do It in Chemical Bonding!*

**Here are some fun, flirty ways to incorporate the designs of chemical bonds into some sexy bedroom bonding time! ;)**

**1.**
Pretend like a soft scarf is the electron cloud. Envelop him in it, then tie him to the bedpost as if he were an oxygen atom and the two posts of the bed were hydrogen. You just made sexy water!!

**2.**
Get on the floor and handcuff his right hand to his right foot, and left hand to left foot. Now he's like *benzene*, a hydrogen and carbon atom arranged in a ring. If you like it then you shoulda put a RING ON IT!

**3.**
A fun safe word is *2-(2-propyl)-5-methyl-cyclohexane-1-ol.* For short, $C_{10}H_{19}OH$, or CCCCCCCCCCHHHH-HHHHHHHHHHHHH-HHOH.

**4.**
Blindfolds are a good way to start bondage "lite." Get a good satin blindfold that won't let any light in. Then stick the blindfold up his butt. It's a good sex place! The butthole! I promise!

# Gases, Gases, Everywhere, but Not a Drop to Stink!

P.U.! What's that? It's gas! Ha, just a little joke for you! Gas doesn't just mean the stuff that comes out of your fetid butthole. It's also the stuff that is in your oven and you can kill yourself with it. It's honestly a beautiful way to die. Like if I found a woman who killed herself in an oven? I'd be like, at least she died doing what she loved: being around ovens. **FIG. 2.13**

Because gas is usually invisible, you might not think about it surrounding you, but when you're naked you're actually being touched by a million fingers of gas. In a lot of ways, gas is like God. Gas is all around you. Gas watches you when you sleep. It embraces you when you undress. They have both in Europe. It's used in wars.

**FIG. 2.13**

## Selfie-a-Plath

*Love you, Sylvia! Miss you, babe!*

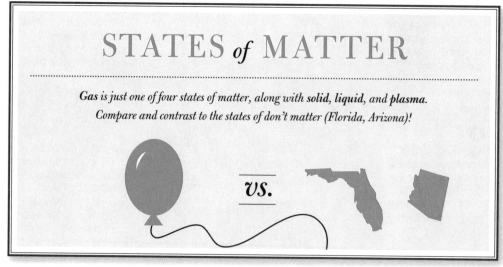

## STATES *of* MATTER

*Gas is just one of four states of matter, along with **solid**, **liquid**, and **plasma**.*
*Compare and contrast to the states of don't matter (Florida, Arizona)!*

*vs.*

# Nuclear Chemistry

Here is some intense stuff, babes! Nuclear chemistry deals with radioactivity. It's like when you look super hot and a guy at the club is like, "Yo, you radioactive, gurl!!" except that radioactivity in real life causes your skin to fall off, cancer, etc. And it has nothing to do with radios, okay, so stop asking if it does or I swear to God I will pull this fucking book over. FIG. 2.14

so freaking pretty!! I know you're not supposed to get close to radiation because of the skin falling off, cancer, etc., but honestly how could anything that glowing and hot be bad? If getting close enough to a glowing, radioactive isotope to touch its mortal-wound-inflicting heat is wrong, I don't want to be right! FIG. 2.15

FIG. 2.14

FIG. 2.15

Certain variations of elements (called *isotopes*) are unstable and naturally break apart, releasing *alpha particle emissions*, *beta particle emissions*, or *gamma radiation emissions*. Almost as unstable as my mom! Mother-daughter relationships are very hard. But who cares about that? Let's get back to the science—radiation is

# MOST EMBARRASSING MOMENTS

"I had spent all day in the radioactive lab and by the end of the day I was HORNY. I was in bed with my boyfriend when my skin started melting off! My time of the month, I guess! FML."
**—Christa, 23**

## Meltdown Edition

"I bought my boyfriend tickets to a hockey game for V-tines Day. We were having a blast, but then the Jumbotron broadcast us on the big screen RIGHT as both my eyes fell out because I had been exposed to intense radiation at my internship! Four-eyes? More like no-eyes, ugh! Well, actually, I still had my glasses on (so cute, Kate Spade) so I guess I was still a two-eyes."
**—Tina, 23**

"I had a huge crush on one of my superiors at Chernobyl, where I worked as a reactor attendant. I was so love-struck that I bumped this one switch that caused reactor number four to have a sudden and unexpected power surge, and when an emergency shutdown was attempted, an exponentially larger spike in power output occurred, which led to a reactor vessel rupture and a series of steam explosions. Thirty-eight people died. And what's even worse? My superior doesn't even know I exist! Because he's dead!"
**—Ivanka, 49**

# Air Pollution

Smog is created through the chemical reaction of *sunlight*, *nitrogen oxides*, *ground-level ozone*, *sulfur dioxide*, and *carbon monoxide*. It is created when smoke from industrial waste mixes in the atmosphere and is especially harmful for senior citizens, children, and people with heart and lung conditions such as emphysema, bronchitis, and asthma.

But it can also be a really beautiful and romantic backdrop to a make-out sesh with your cute guy (or gal, lesbo!). Here are some of the most romantic places to see smog.

## Most Romantic Places to See Smog

"In London, when the sun is coming up. I love breathing it with my boy Sam!"
—**Sally, 19**

"In front of the Blue Mosque in Istanbul with my boyfriend, Simonn."
—**Marta, 23**

"One word: Port-Au-Prince. Is that one word? Three words? Whatever! It's so cute!"
—**Silvelia, 21**

"Shanghai. Or maybe this is Beijing? I can't see very well with all this smog."
—**Cough, cough**

"You can really see it for miles from this adorable art park in Los Angeles. My boyfriend, Simonn, loves to picnic there."
—**Kelsey, 28**

"On top of Machu Picchu."
—**A Peruvian girl,** *~3 in Incan years (28 in human years)*

"Wait, what did Kelsey say? She has a boyfriend named 'Simonn' with two *n*'s??? There are definitely not two Simonns, that fucking bitch." —**Marta, 23**

"That's right, cunt, your boyfriend Simonn is two-timing you. One timing for each *n* in his name."
—**Kelsey, 28**

"I'm going to fucking kill you, bitch, and Simonn too."
—**Marta, 23**

"Hey, guys, one sec . . . 'Sam' is just a nickname for my boyfriend . . . I think his real name is . . . Simonn . . . "
—**Sally, 19**

"I love love!"
—**Simonn** (deceased)

R.I.P. Simonn

# Famous Chemists

*Radon* and *X-rays* were discovered by Marie Curie. You could say this next section is *X-ray-ted!* Because she invented X-rays, and because here's a peen!

## Marie CURIE *vs. Marie* CLAIRE

**In this piece, I will pit Marie Curie, possibly the greatest female scientist of all time, against *Marie Claire*, a women's magazine that recently featured a ripped, shirtless Justin Timberlake on the cover. It's anyone's game!**

**MARIE CURIE**
Marie Curie was the first woman to win a Nobel Prize, the first person to win two Nobel Prizes, the only woman to win in two fields, and the only person to win in multiple sciences.

*MARIE CLAIRE*
Has a thing about hairstyles!

—

**MARIE CURIE**
In 1995, she became the first woman to be entombed on her own merits in the Panthéon, Paris. The curie (symbol Ci), a unit of radioactivity, is named in honor of her and her husband, Pierre. The element with atomic number 96 was named curium.

*MARIE CLAIRE*
Has a thing about punk-rock nail art!

**MARIE CURIE**
Probably had a bunch of burns on her face from the radiation she was exposed to while discovering radium. Probably didn't even have the right-shade concealer to hide her radiation burns.

*MARIE CLAIRE*
Even though she's a magazine and doesn't technically have a face, she would never leave the house without looking impeccable.

—

**MARIE CURIE**
Died from radiation poisoning.

*MARIE CLAIRE*
Can't die and will continue to put out magazines FOREVER!!!!

—

**MARIE CURIE**
A 7 in the right light. The right light does NOT include the glow coming from the radon she kept in her desk drawer. She was a dog in her radiation lab. Like, a 5, at best.

*MARIE CLAIRE*
Has a thing about beach-ready looks!

**MARIE CURIE**
Real butterface. Because of the radiation burns. Real butterradiationburnsonherface.

*MARIE CLAIRE*
Only weighs like a pound. At the most three pounds in September. Lucky!!

—

**MARIE CURIE**
Didn't know who Justin Timberlake was. Like some sort of lesbo. Butterfaceisfallingoff.

*MARIE CLAIRE*
Had him on the cover! And he was RIPPED!!!!!!!!!!! Yo, Justin, are you the condom we use when we have sex and I try to trap you forever by secretly getting pregnant? Because you are SECRETLY RIPPED.

**WINNER?**
***MARIE CLAIRE!!!!!!!***
Obv. *Marie Claire* is my number-one role model and best friend. Sorry, Marie Curie!! Maybe try winning a Nobel Prize in being a BUTTERFACE!!!!

# Chemistry Recap

How fun was THAT? No actually, I'm asking. I really don't know what's fun or not anymore. I've totally lost all concept of "fun." But that was fun, right?

And, P.S., I've been doing some thinking over these few months (but never while I'm driving, don't worry!). I think I'm ready to end this long-distance relationship with Xander. I've even slept with a select few of my favorite football teams to ease the transition! So it looks like I'm back on the market, ladies!

# Recap Questions

**QUESTION 1:** How quickly did you come in the Bucking Bronco position?

**ASSERTION 1:** That was a trick question, women can't come!! (More on this later!)

**QUESTION 2:** How do you get vomit off a chicken bone?

**QUESTION 3:** Has your menstrual cycle synched with mine yet?

**QUESTION 4:** If not, why not? Do you hate me? Are we not best friends anymore? Because I was about to invite you to meet my mom. We were going to have a *Clueless*-themed party and all watch *Clueless* together.

**QUESTION 5:** Did question 4 count as one or three questions?

**ASSERTION 2:** I love you girls even more than I ever thought I could. :)

# 3 Physics

LET'S GET PHYSICS, Y'ALL!!!!!!!!!!!!!!!!! SEE—I REMEMBERED!!!!!!

**LET'S GET PHYSICS, Y'ALL!**
(A PARODY OF "PHYSICAL" BY OLIVIA NEWTON-JOHN)
I'M SAYING ALL THE THINGS THAT I KNOW YOU'LL LIKE
ABOUT PHYSICS AND SCIENCE
I GOTTA TEACH YOU FACTS JUST RIGHT
YOU KNOW WHAT I MEAN! ;) (PERFORMANCE NOTE: For the winking face, you can either actually wink or sing the words "winking emoticon.")

I TOOK YOU TO A CHAPTER ON BIOLOGY
AND THEN A REAL CHEMISTRY CHAPTER
THERE'S NOTHING LEFT TO TALK ABOUT
UNLESS IT'S PHYSICS OR ALL THE OTHER CHAPTERS

LET'S GET PHYSICS, Y'ALL
PHYSICS, Y'ALL
I WANNA GET PHYSICS, Y'ALL
LET'S LEARN ABOUT INCLINED PLANES, INCLINED PLANES
LET'S LEARN ABOUT INCLINED PLANES

# Introduction

I think we all agree, that song that I wrote was amazing, gals! So, physics is kind of difficult to teach because it's not just a soft science like bio and chem, it's a super *hard science*. Usually I like when things are hard (Can I get a "dick as hard as a diamond and as red as blood, aka a blood diamond"?!), but when it comes to sciences, I like them soft and flaccid, like my boyfriend when I showed him my "twin."

Physics comes from the Greek φυσική (ἐπιστήμη), which, loosely translated, means "illegible." Physics looks at matter through space and time. So, while chemistry studies the mixing of matter, physics studies how matter moves and exists.

Physics is probably the most important science. There are international committees that are charged with making advances in physics. You'd think we as a human race would figure out that there are more important things to spend money on than physics, but I guess not! All that money could be going toward figuring out more kewl things to top frozen yogurt with (NOTE TO SELF: try regular yogurt???), or why you could be a grown woman and your titties could still look like two little mosquito bites, *Lauren*. Lauren is my best friend who has the tiniest titties. They basically look like two Girl Scout merit badges, except that it's the opposite of merit. It's like her chest is a Girl Scout vest and you get badges for worthlessness and she got both the worthlessness badges you get when you have tiny boobs. Boom, Lauren. You got *burned*.

And guess who also got burned? Xander, for being a little bitch! I've finally moved on! I know you were worried I wasn't going to get over him so easily, but I am doing great, girls. In the time since I last wrote, I've slept with every member of Xander's ninth-grade hockey team! I am SO OVER HIM!! Over him like his old high school hockey friends were OVER ME as they GANGBANGED ME IN THE PARKING LOT OF AN OUTBACK STEAKHOUSE!!

## Math Review

Before we get into the meat of physics, we gotta review some basic math. Physics is very math-heavy, so if you don't know math, this is not going to be a fun Wednesday night for you! You might as well break out the Wednesday wine and give it all up!

Let's quickly cover math basics. Here's a mnemonic device: 4 before 5 except after 6! Also, remember this little joke to figure out the order of numbers: Why was 6 afraid of 7? Because 7 8 9! And why was 9 afraid of 10? Because 10 is a Crip-affiliated black man!

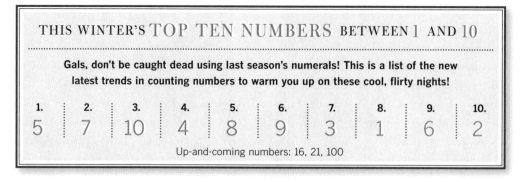

THIS WINTER'S TOP TEN NUMBERS BETWEEN 1 AND 10

Gals, don't be caught dead using last season's numerals! This is a list of the new latest trends in counting numbers to warm you up on these cool, flirty nights!

| 1. | 2. | 3. | 4. | 5. | 6. | 7. | 8. | 9. | 10. |
|----|----|----|----|----|----|----|----|----|-----|
| 5  | 7  | 10 | 4  | 8  | 9  | 3  | 1  | 6  | 2   |

Up-and-coming numbers: 16, 21, 100

Well, there we go—that's pretty much all of math! Congrats!

# Why, Scientifically, Women Can't Drive... for Her!

As the old saying goes, women are like snow-flakes: they can't drive. This is a part of science that I feel very passionate about. If I can save even one life of a lady driver or pedestrian (male or female), this whole book-and-writing ensemble will have been worth it.

Fashion check-in time! Right now I'm wearing a white cotton T-shirt that has the left boob cut out, an eye patch over my left nipple, and the bottom half of one of those huge cakes a stripper is supposed to pop out of. I tried to pop out of one for my current boyfriend Anton's birthday (he played goalie), but I think I've put on a little weight in the hips and I couldn't quite get out. :( I hate that I can't still fit into my stripper cakes from high school.

We've all been there: you're in your cute little car, singing along to "Firework" by Katy Perry, and suddenly you've mowed down six pedestrians and a family of geese **FIG. 3.1** like they were blackheads and you were a super good Bioré blackhead remover strip (SPONSORED TWEET).

Here's the thing: it's not your fault! Due to your inherently poor grasp of physics, there is no reason that you *should* be able to drive.

**FIG. 3.1**

---

### REASONS SCIENTIFICALLY
### *Why Women Can't Drive*

$$\text{Speed} = \frac{\text{Distance}}{\text{Time}}$$

**SPEED**

*Speed* is how much distance is covered over a certain amount of time. Women can't control the speed of their cars because speed is measured in numbers, which are a major aspect of mathematics, something that is immediately erased from your memory when you listen to Katy Perry.

$$\text{Velocity} = \frac{\text{Speed}}{\text{Time}}$$

**VELOCITY**

*Velocity* is speed over time. It measures a rate of change of speed. Women can't control the velocity of their cars because their pocket mirrors/gems/Greek yogurt fall onto the accelerator pedal, making the car speed up infinitely until the car explodes/they reach the mall.

$$d = vt + \frac{1}{2} at^2$$

**DISPLACEMENT**

*Displacement* is the distance between an object's initial position and its final position. Women can't understand the displacement of their cars because it's measured in feet and feet reminds us of shoes (Louboutin!) and then we forget how many feet we're supposed to be thinking of!

**SPATIAL REASONING**

In general, women have no idea where their bodies or cars are in space. Right before I wrote this, I ran into a clean window just like a bird! I guess that's on me for cleaning my windows so gosh darn well every Sunday!

# LADY DRIVERS!

## *Can't live with 'em . . . !*

Women should be encouraged to do all they can to push back against their shortcomings. *Science . . . for Her!* held a contest to find the best female driver in America! Lady drivers were judged on how many accidents they'd been in, how many fatalities per accident, and number of driving gloves owned (pairs not singles), as well as other categories. Entrants also had to write an essay explaining why they were a good female driver despite their inborn shortcomings, and why any pedestrian fatalities were the victims' faults.

After taking all the parameters into consideration, we finally picked a winner—Carly Jansen, twenty-nine, of Bangor, Maine!! Unfortunately, Carly was killed in an auto accident as she was answering her cell-phone call from us telling her she'd won, so we moved on to the runner-up, Lizzie Lambert, thirty-three, of Ann Arbor, Michigan! She's pretty good, too.

# The
# TOP WOMAN DRIVER
## *in America*

**Name:** Lizzie Lambert

**Age:** 33

**Hometown:** Ann Arbor, Michigan

**Number of accidents:** 2

**Number of fatalities:** 1

**LIZZIE'S ESSAY:** Even though I was born in the prison that is a woman's body, I feel that I have worked night and day to overcome this living hell of a gender. I played a lot of pool to practice spatial reasoning. I invested in conservative footwear (two-inch stilettos with studs on the cuff) so that my foot would rarely slip off the brake and onto the gas. I took two extra sessions of driver's ed so that I would be fully prepared to merge on AND off the freeway. I made a real commitment to putting on my makeup BEFORE I started driving, and if I had to put on makeup while driving I waited for long stretches of straight road so I could safely take both hands off the wheel. I realized that "Do Not Enter" signs were not suggestions, but should be respected as canon. I removed enough of the "Proud Mother of an Honor Student at Rushfield Middle School" bumper stickers that I had plastered the windshield with so that there were two eyeholes to look out through. I removed all the stuffed animals from the back window except the Spongebob because he's so cute!!!!! I've stopped sticking my head out the moonroof as a hair dryer. I started holding the steering wheel at 10 and 2. I learned to read an analog clock so I could figure out what "10 and 2" means (so retro!). I've stopped making my honor student at Rushfield Middle apply my lipstick for me. I've stopped tailgating so that I could check out the hair extensions of the gal driver in front of me.

**LIZZIE'S EXPLANATION OF HER FATALITY:** I unfortunately totaled my car in my first driver's-ed course, but the driving school that I was taking it from was super classy. They took full responsibility for not teaching me how to merge and, even though I killed my driver's-ed instructor, they gave me a coupon for two free classes. I'm writing this essay during the second one! Shut up, Mr. Know-It-All teacher, I do what I want!

*While giving this interview, Lizzie crashed her car into a car containing the second-, third-, fourth-, and fifth-best remaining woman drivers in America. There were no survivors, and no other women in America know how to drive. Better luck next year!*

# Atoms

A lot of physics deals with li'l atoms and how they split apart and squish together. When atoms split or join, they unleash a lot of energy in the form of physics. See, a lot of good things come in small packages! That's what I used to say about my ex Karl, who had a package that was so tiny that I sometimes used heat-finding night goggles to see where it was. Turns out he didn't have one. Karl was a woman. I should have known when he kept telling me his name was Carla and he was my roommate at our all-girls school. Isn't it annoying how all the good guys are either taken or women?

But if you're a normal red-blooded lady like myself (Type AB, baby! I'm the universal recipient . . . of PENIS! ;) Can I get a "Go to www.redcrossblood.org to find out how you can donate blood to your local Red Cross Blood Bank"?!), the word *atom* just reminds you of some sexy-ass Adams that you want to jumpity-hump.

## Atoms *vs.* Adams

**HYDROGEN ATOM VS. ADAM LEVINE**
Adam Levine has crazy sex bones and tats. The hydrogen atom is small and trim, for those of you ladies who enjoy good things that come (cum!!) in small packages, like I said.

*WINNER: Adam Levine*

 vs.

**NEON ATOM VS. ADAM BRODY**
Adam Brody combines the handsomeness of a symmetrical face with the whimsy of being a Jew. In recent years, the neon atom has really let itself go.

*WINNER: Adam Brody*

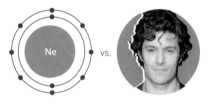

**NITROGEN ATOM VS. ADAM LAMBERT**
Adam Lambert is a gay icon. So is nitrogen!

*WINNER: Nitrogen atom*

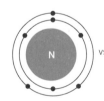 vs.

# Get Down on All "Force" and Stick Your Finger in My Butthole

Fyi ladisdae I amm weaibrsg tghe CUTEAST glaovse rigsht noow!!!!!!!!!!!!!! Ganna takea ehm off ni a ssec.

Okay, gloves are off. ;) Sorry, they were so cute. Juicy Couture. A *force* is any influence that causes an object to undergo a certain change, either concerning its movement, direction, or construction. Force is in your life without you even thinking about it.

That means you have more space in your tiny brain to think about important stuff, like Louis Vuitton sushi mats!

You unknowingly utilize force when you reheat a Hot Pocket, flutter your eyelashes at your new crush, or "Reheat a Hot Pocket" (this chapter's sex move; rated "black diamond" for difficulty).

CHECK IT OUT!

# PHYSICS *through* NAIL ART

**Physics is so much more fun and understandable when it's presented on cute nails!**

### Converting Tangential Acceleration to Angular Acceleration

You'll never forget how to convert tangential acceleration to angular acceleration with this cute getup! #nails #cute #tangentialacceleration toangularacceleration #pretty

### Centripetal Force through a Curve

To get the "V" just right, use a nail-polish pen and clean up the straight lines with some nail-polish remover on a Q-tip (SPONSORED TWEET). If it's a shaky line, you might think the equation is "$Fc=(m)(U^2/r)$" and that logic hole will totally fuck up your outfit! #cute #fun #logicholes #fuck

### Equation of Mass-Energy Equivalence

The kittens, Barts, etc. aren't part of the equation but they add a little pizzazz to this VERY BORING equation that everyone already knows! Don't forget—THERE ARE NO BARTS IN THE EQUATION. If you put Barts in the math, it won't make any sense. Bart Simpson is not a number! An easy mnemonic to remember that Bart Simpson is not a number? "BSINANBIY-EMTMAUBSAANIMAWSLYABYBFF" ("Bart Simpson Is Not A Number But If You Ever Make That Mistake And Use Bart Simpson As A Number, I, Megan Amram, Will Still Love You And Be Your Best Friend Forever"). You'll never make that mistake again! But if you do, I'll still be your best friend!

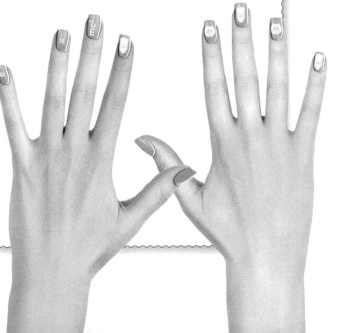

# Weight *Laws* Tips, Get It, Instead of "Weight Loss Tips"!

I know we've flirted around the idea of weight in this book so far but this is finally going to look at the *science* behind why your ex–best friend Katja's ass has ballooned up like a rotten jack-o'-lantern.

*Mass* is different from weight. But both of them are BAD. Mass is the actual amount of stuff something is made of, and *weight* is how much that mass weighs in a certain gravitational field. So like, if you weigh 120 pounds on Earth (and I do!! JK I weigh way less!!!), that doesn't mean you'd weigh that much in different gravitational fields. In fact, Earth is one of the worst places to be if you're a 120-pound heifer.

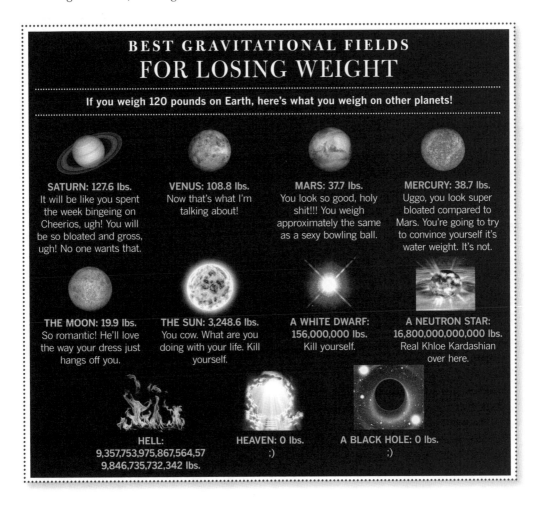

**BEST GRAVITATIONAL FIELDS**
**FOR LOSING WEIGHT**

If you weigh 120 pounds on Earth, here's what you weigh on other planets!

**SATURN: 127.6 lbs.**
It will be like you spent the week bingeing on Cheerios, ugh! You will be so bloated and gross, ugh! No one wants that.

**VENUS: 108.8 lbs.**
Now that's what I'm talking about!

**MARS: 37.7 lbs.**
You look so good, holy shit!!! You weigh approximately the same as a sexy bowling ball.

**MERCURY: 38.7 lbs.**
Uggo, you look super bloated compared to Mars. You're going to try to convince yourself it's water weight. It's not.

**THE MOON: 19.9 lbs.**
So romantic! He'll love the way your dress just hangs off you.

**THE SUN: 3,248.6 lbs.**
You cow. What are you doing with your life. Kill yourself.

**A WHITE DWARF:**
**156,000,000 lbs.**
Kill yourself.

**A NEUTRON STAR:**
**16,800,000,000,000 lbs.**
Real Khloe Kardashian over here.

**HELL:**
9,357,753,975,867,564,57
9,846,735,732,342 lbs.

**HEAVEN: 0 lbs.**
;)

**A BLACK HOLE: 0 lbs.**
;)

# Heating Up & Cooling Down

*Thermodynamics* is the study of how heat is transferred from one thing to another. *Dark matter* is good at absorbing heat. Um, speaking of "dark matter?"—I *love* black men! ;)  Sorry to be TMI but they are *great* lovers! I got bit by the jungle fever bug (*malaria*) when I was like thirteen and I have not looked back.

FIG. 3.3

FIG. 3.2

## THREE FUN WAYS TO ENJOY A DATE

*Even If You Are White and Your Date Isn't the Same Race as You*

**1**

Go to a movie that involves lots of races! Try *Birth of a Nation* or, if you want to spend a night in, rent *Roots*.

**2**

Eat food that can reach both of your cultures! Force yourself to scarf down *fried chicken*, an exotic delicacy the black people stole from Colonel Sanders in 1642!

**3**

Get into a little sexy role-play that helps you understand where they're coming from! ;) Make your black beau pretend to be Sally Hemings, a colonial black slave, and you can be Thomas Jefferson, who raped her and fathered one or more of her children. Then you'll *truly* know where Sally Hemings was coming from—her black penis!

# Science . . . for Urban Her!

Let's take this opportunity to talk a little bit about race, babes! Here's a little racial humor to get us warmed up: isn't it funny how white people read textbooks like *this*, while black people read textbooks like *this*? This is not a very good joke written out!

As we've established, humans are animals, just like the animals of the sea and sky. Therefore, differences between *races* are literally exactly like breeds of dogs, if one paler breed of dog had the proclivity to hold other slightly darker dogs captive!! Slavery is mega-bad but OMG—how cute would that be? :) A white poodle forcing a chocolate lab to pick his dog-cotton! OMG so cute! I think it would look a little something like this! ;)

*12 Years a Slave?* More like *84 DOG YEARS A SLAVE!!* But yeah, slavery is mega-bad!

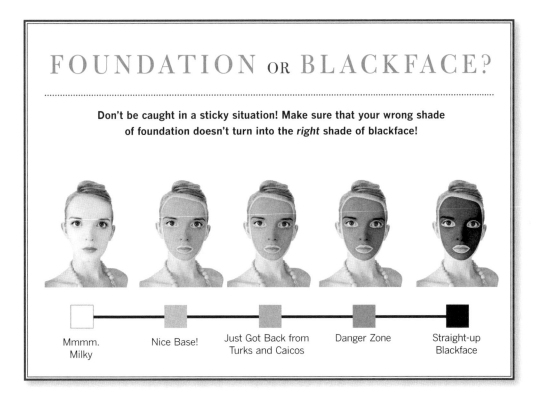

FOUNDATION OR BLACKFACE?

Don't be caught in a sticky situation! Make sure that your wrong shade of foundation doesn't turn into the *right* shade of blackface!

Mmmm. Milky | Nice Base! | Just Got Back from Turks and Caicos | Danger Zone | Straight-up Blackface

# Laws of Physics . . . *Jude* Laws of Physics! I'm Talkin' Hotties of Physics! ;)

There are laws for physics, just like there are laws for how many carbs you're supposed to eat. First law of both physics and carbs is "EAT NO CARBS." FIG. 3.4

Sir Isaac Newton (9 body, 4 face) invented *Three Laws of Physics*. But, hey, laws, shmaws.

Physics should be done by following the *heart*. Here are some of the most eligible bachelors of physicsdom! (Dead people count as bachelors.) FIG. 3.5

FIG. 3.4

FIG. 3.5

# BACHELORS *of* SCIENCE

**GALILEO GALILEI**
(primitive astrophysics)
Galileo was an Italian physicist, mathematician, astronomer, and philosopher, but he also spoke the universal languages of love and Italian. He is best remembered for the Galileo, a sex move where the man enters the woman as both are falling at an equal gravitational rate from the Leaning Tower of Pisa. (This chapter's bonus sex move!)

**ISAAC NEWTON**
(classical mechanics/gravitation)
Sir Isaac Newton, an all-around science genius, was alive from 1643 to 1726, which means that his lifetime spanned the year 1690, the ultimate 69. By all accounts, he spent the year 1690 giving oral pleasure to several British babes at once. He is famous for the three laws of sex, including: a penis in motion tends to *stay in motion*. OOOOOH GIRL!!

**AMEDEO AVOGADRO**
(microscopic physics)
Avogadro is most well-known for *Avogadro's number*, approximately $6.022 \times 10^{23}$, which is the number of women he CROTCH-MALLETED (rounded to the nearest trillion)!!!! If you wrote down all the names of all the women Avogadro down-there-mistletoe'd, it would stretch to the moon and back thirty million times (rounded to the nearest thirty million).

**ALBERT EINSTEIN**
(special and general relativity)
Though many know him as one of the most famous scientists who ever lived, Einstein was also a notorious playboy who invented the condom because he loved to pork but didn't want any STDs. *E = MC squared? U = VD spared!*

**ERWIN SCHRÖDINGER**
(quantum mechanics)
Famous for the hypothetical "Schrödinger's cat." Which is referring to his mistress Marta Schrödinger's pussy!

I would date any of them! Because unfortunately I'm single again. During the writing of this chapter my hockey boyfriend Anton dumped me, saying that I was tearing him and his hockey teammates apart and the "Brotherhood of the Ice" was more important than any girl. Joke's on him: I replaced his puck with a used diaphragm!

America has historically been good at pro-ducing physicists. Many famous physicists have come from America: *Benjamin Franklin*, *Robert Millikan*, *Dr. Pepper*. The soda was called "Mr. Pepper" before he got his PhD in nuclear physics. However, not everything in America's past has been as successful. America is a very complicated place. I think this review that I found online in a blog pretty much sums it up.

# AMERICA

## ★ ★ *A Review* ★ ★

How to begin this review? Few countries that debuted in the 1700s have been as controversial or long running (it's into its 236th season now) as *America*. It may not have the staying power of perennial favorites such as *China* or the credibility of indie darlings like *Finland*, but America has proven that it can at least make some cultural impact. It's not the best, but hey, they can't all be *Louie*.

America was originally a spin-off of the long-running *England*. Airing from the 1776–77 season through today, *America* focuses on a small ensemble of white people using things in the ground to become rich or kill brown people. A sprawling dramedy, it combines all of the loose plot points of a Tyler Perry sitcom with all the fun of being white.

It has widely focused on the themes of war, freedom, sitting, Fenway Park, maps, the one true Christian god, rugs, pregnancy tits, *VICE* magazine, butterfaces, coal, butterdicks, "Where's the Beef?," Chicago, Larry Flynt, colonialism, Terri Schiavo, NBC single-camera sitcoms, toddlers, suicide pacts, Atari, penny-farthing bicycles, SpaghettiO's (Cool Ranch flavor), tiny dolls, the TLC show *Sister Wives*, H1N1, television, and genocide. It has some unique perspective every once in a while, but honestly, *America* can be super derivative. Most of the stories have already been on *The Simpsons*.

A lot of episodes in *America* don't really hold up. Slavery? Parachute pants? White slavery? It just feels really overdone now. Among the most memorable episodes are "The Civil War," "Texas," "World War" (a two-parter), and "Black President."

Some of the story lines are also a bit of a stretch. Are they really expecting us to believe that they killed *all* the Indians and that all those Indians did to deserve it was invent diabetes?! And come on—that stuff in the 9/11 episode could not have happened without someone working on the inside. That makes no sense. "9/11" jumped the shark. *Hard.*

It's been on so long that no one wants to comment on the OBVIOUS PLOT HOLES. Such awful continuity. Like, how could it be explained that in season 170, George H. W. Bush fathered a retarded son, but then in season 225, that son became president?! Really terrible continuity. I would like to point out that I appreciate a recent callback to earlier plots. Around seasons 174–184, some of the anti-feminist and sexist story lines were put on the back burner, but it's nice that we've seen a resurgence in this last season.

There's a lot of homosexual undertones to the country. The Very Special Episode about Lewis and Clark was revised to not include the fact that they originally named Oregon after the French word for "gay-ween butt-orgy" ("*Baguette*"). Baseball, the "American Pastime," is about using bats ("dicks") to hit balls ("balls") all while blowing each other in the dugout ("RBI"). And how about the American flag? Obviously thirteen dicks going into fifty buttholes.

*America* has time and time again proved itself as a launching ground for young starlets. It's fun seeing people before they became huge stars, like John Ritter, Stella McCartney, Theodore John "Ted" Kaczynski, and Ted "Ted" Bundy. But the ensemble works best when we see the regulars yearn for a raise or promotion, struggle with Mary Tyler Moore's foibles, and be there for Mary Tyler Moore when the going gets rough. I stole this from a review for *The Mary Tyler Moore Show*, but I think it completely and entirely makes sense to literally lift from that review and drop it into this context as well.

As someone with more quirky and alt tastes, I can't say that *America* is my favorite thing to watch. I'm more into *Breaking Bad*. Have you seen season four?! Season four of *Breaking Bad* is flawless. Season four of *America* is VERY uneven. It had no main black characters. *Girls*, much?! I love *The Wire*!

I just hope to God (the American/right one) that they don't pull some deus ex machina shit at the end of this series. Like, there's nuclear war with North Korea, or they've been dead the whole time or something.

Anyway, it may have veered off wildly from the pilot, but *America* is definitely worth a look. It's an interesting experiment in the world of prime-time sovereign nations. What the characters lack in consistency, they make up for in body weight, lingering racism, and inconsistency. But it makes for a quick and easy viewing, and can often surprise you with heartfelt turns. It's like eating Cool Ranch SpaghettiO's on a warm summer's eve. And hey, sometimes things get really good right before they're canceled.

MY RATING: 50 stars (out of 100)

★★★★★★★★★★★★★★★★★★★★★★★★★★★★★★
★★★★★★★★★★★★★★★★★★★★★★★★★★★★
★★★★★★★★★★★★★★★★★★★★★★★★★★★★★★
★★★★★★★★★★

# Flat Planes, Or: My Best Friend Lauren Who Wears a 32A Bra

You're flat like Lauren? Oof magoof, girl, this is a real sticking point. The laws of physics basically state that if there's something (e.g., a hand or peenium) traveling along a straight line and it meets a flat curve (e.g., your flat seven-year-old-boy chest), it won't have any friction and will just fall off your body. Then your boyfriend or hookup partner might even hurt himself if he's traveling fast enough. He might slip off your itty bitty titties and fall on the floor onto a knife or something.

I want to reiterate what I said about Lauren—her breasts were so small that they were basically concave. They looked like two strawberry cough drops from Halls (SPONSORED TWEET).

That being said, some women don't like their big breasts. There's something called a "breast reduction," which can reduce your big melons into tiny melon balls. I'm starting a charity to help women with very large bingos (like myself) help out less fortunate uggos like Lauren (like Lauren). You donate your extra boob material to mosquito gals who are unlucky or poor enough to have tiny boobs. It's called Yabbos of Love and it will be up and running as soon as I sleep with the entire Better Business Bureau.

> **Good Name for a Vagina I Just Thought of**
> "Better Business Bureau"

# The Future of Physics

The physics I've told you about is pretty cool so far, right? Well, it's nothing compared to the future of physics. Part of the future of physics is time travel. Using the laws of matter and motion, science is soon going to be able to transport your big summer booty (better lay off the Better Cheddars!) to any time or place you desire (Try the winter, when you don't need that bikini body, you fat woman! You're just a big fat lady! Pudge-a-pudge-a-pudge-a! Buffalo Bill in *Silence of the*

*Lambs* would have wanted to skin you because you're fat and have a lot of skin to go around!).

As a side note, I just want to preface this by saying time travel very likely already exists. I mean, there's no way to prove that all murders aren't just time travelers killing future Hitlers. I truly believe that many scientific advances such as time travel have already been made and the government is just totally covering it up.

# THIS SPRING'S TOP COVER-UPS!

**There are so many great cover-ups out there! There are also a lot of bad cover-ups that the American government has pulled that might fool most of the goddamn sheeple in this country but not me, you better believe that I have not had the wool pulled over my eyes! I will now list the top cover-ups for this fun, flirty season!**

**8.**

### FLIRTY SARONG
Boxy, flirty, and SUPER comfy with whatever you're wearing! Added bonus: it's got a flower on it! Cute!

**7.**

### MOON LANDING
The moon landing in 1969 was a hoax staged by NASA that the US government covered up by visibly doctoring the photos of the fake staged moon landings on soundstages in Hollywood, California, killing off astronauts and other affiliated people who knew the truth about the moon landings, and disposing of blueprints and recording devices that would have proved that America had never landed on the moon. Cute!

**6.**

### CHIC CARDI
Want to conceal your trouble spots? Throw on this floaty, flirty light sweater! Cute!

**5.**

### MARTIN LUTHER KING JUNIOR'S ASSASSINATION BY LYNDON B. JOHNSON
It is widely known that MLK Jr.'s assassin, James Earl Ray, was just acting as a fall guy in an epic conspiracy that included President Lyndon Johnson, the Joint Chiefs of Staff, J. Edgar Hoover, and several army units as well as organized-crime figures. But the government has tried to convince us that James Earl Ray was working alone and that they wouldn't be interested in picking off one individual black man—cute!

**4.**

### HOT HOODIE
This hot hoodie puts the *hot* in *super hot*! Cute!

**3.**

### JFK ASSASSINATION
The John F. Kennedy assassination of 1963 is widely regarded as the biggest US government cover-up that has ever been attempted. Though Lee Harvey Oswald was apprehended as the "single" assassin of the president, JFK was truly murdered due to a wide-reaching conspiracy that involved the CIA, the KGB, FBI director J. Edgar Hoover, sitting vice president Lyndon B. Johnson, Cuban president Fidel Castro, and anti-Castro Cuban exile groups. Cute!

**2.**

### SEE-THROUGH SCARF
Lets the beach boys see just enough sexy curves! Cute!

**1.**

### TIE BETWEEN THE LBD (LITTLE BLACK DRESS) AND 9/11
The US government managed to cover up the fact that the World Trade Center towers' iron could never have burned at the temperatures that the government says it did or else it would have never left nano-size iron balls in the dust that clearly show that the towers burned at a temperature consistent with a controlled inside burn. Additionally, according to the 9/11 Commission Report, the cockpit "black boxes" from flights 11 and 175 were not recovered from the remains of the WTC attack; however, two men who worked extensively in the wreckage of the towers said that they helped federal agents find three of the four black boxes from the jetliners. Other option: the LBD, which flatters any body shape or size. Cute!

Whether or not the government has secretly been using time travel already as its own fun little secret (ooh, "Fun Little Secret," another good name for a vagina!), you're going to want to jump on this bandwagon ASAP as soon as it's available.

Time travel is the perfect accessory for us ladies. First of all, time travel is a great way to lose weight. If you travel back in time to when you were a baby, you'll be like a size negative 6! Also, Xander won't have broken up with you if you're a baby. And then you can rewin him back. Because you'll be as skinny as a baby. Tell Xander he can touch the soft spot on your head with his wiener or something.

Plus, think of all the fun things that will have been invented to buy and do in the *future*!

## New Year's Resolutions *for* Year 3014

 **1**

Lose fifteen pounds from your problem areas (hips, space-boobs, vestigial face)

 **2**

Spend more time with your government-rationed .452 of a son or daughter

 **3**

Take the family on a trip to www.nature.com

 **4**

Volunteer at your local chapter of the White People Remembrance League (white people have been extinct since 2021; you are an exotic mixture of brown and Asian and Google Glass)

 **5**

Pray to the Mother Goddess Zooey Deschanel, who first displayed her omnipotent god powers at the 2016 People's Choice Awards by raising Eleanor Roosevelt from the dead and giving her bangs

 **6**

Learn moon-French

 **7**

Vote for Zooey Deschanel in the 3012 People's Choice Awards as "Best Deity," "Only Deity," and "~*~Kewlest~*~ Bangs"

 **8**

Buy a new Moon Bounce (here on the moon we just call them "Bounces")

**9**

Get promoted from "slave to Zooey Deschanel" to "human sacrifice to Zooey Deschanel" (lateral promotion)

 **10**

Organize your thirty-seven space-boobs by type (normal, brown, formal, or Chicago style)

I just want to apologize quickly. I don't know if you've noticed, but I feel like I've been really judgmental of your weight all through this chapter so far. I think I must be projecting, because I've actually gained more than one hundred pounds over the writing of this chapter.  FIG. 3.6

I think the Xander stuff is finally *really* hitting me, and I've been eating a lot. I've eaten like thirty of those stripper cakes, including the one I couldn't pop out of. Whenever someone would ask me to jump out of a stripper cake, I'd just slowly eat my way out until there was nothing left. I'm big enough (pun intended ;)) to admit that I've been a bad best friend. Love you, gals!!!

FIG. 3.6

# Television

Since TV is basically a box full of magnets, TV totally belongs in the physics chapter. We ladies in America sometimes take TV for granted. Not everyone has a TV, you know. Squirrels are basically TV for homeless people. FIG. 3.7 Every woman should realize that a TV is everything to her. It's going to help raise your children, so you NEED to invest in a good model. It was also most likely the most consistent father figure you had in your life. Know that any issues you have in your dating life are due to the neglect of your TV-daddy.

FIG. 3.7

# *Upcoming*
# TV SCHEDULE

**Get ready for these great upcoming new shows on your favorite cable TV channels in the future!**

| Fri 4/21/31 | 8am | 8:15am | 8:23am | 8:25am | 8:30am | 8:42am | 8:45am |
|---|---|---|---|---|---|---|---|
| **TLC** | Short AND Fat! | Large Cupcakes!: Cupcakes the Size of Full-Sized Cakes! | | Too Many Kids! | Uh-Oh, I Taxidermied My Stepson! | | Not Enough Kids! |
| **MTV** | 16 and Barren | Made: I Want to Have Anemia | Thanks for Nothing, Mom and Dad: Punch Your Dumb Parents with Your Hands if They Don't Get It | | Pimp My Premie (hour-long cleft-lip special) | Pimp My Acai Berry (miniseries) | Pimp My Pimp |
| **THE WEATHER CHANNEL** | Weather Forecast (reruns from 2010) | Some Clouds or Something Whatever | Come Rain or Come Shine: The Chip Milggnman Story | | Whatever (miniseries) | Pimp My Weather | Lava: Not Weather at All |
| **SPIKE** | Face Punch Beach House | Video Game Mardi Gras Beach House | 1,001 Ways to Get Someone Un-Pregnant | Eat This Fart, Guy (miniseries) | Hit in the Balls, Hit in the Nuts (Beach House) | Top Ten Faces to Get Punched in the Face | |
| **OXYGEN** | 1,001 Ways to Make Your Hair Smell Like a Storm | | How to Fart More | Candles and You | Candles or You | Husbands, Husbands, HUSBANDS | Milk Me |

82

# Let's Get Physics, Y'All! Recap

Boy am I TIRED from how much I TRULY LOVED teaching you about physics, and from all the extra weight I'm now carrying around!! I am just so exhausted from enjoying teaching and from imagining how much you loved this chapter (can I get a "I truly enjoyed this physics chapter, Megan, here's my [insert your own credit card number and the little PIN on the back and the expiration date and then record yourself saying it and send it to the publisher's address on the copyright page of this book]"?) I hope you had fun. I'm going to brb and lose all the weight—I can't wait to see you in the next chapter! You won't even recognize me!

# Recap Questions

**QUESTION 1:** Why are my breasts more tender than normal?

**QUESTION 2:** Oh my God, am I pregnant????

**QUESTION 3:** If 9/11 wasn't an inside job, then why do major figures that have dealt with 9/11 investigations keep trying to make public statements about the doctored evidence but they're always silenced by the mainstream media?

**QUESTION 4:** Which makes more sense: that one individual Muslim terrorist was able to choreograph an elaborate dance of planes that was able to bypass every single aspect of US security, OR that the collapse of the Twin Towers and 7 World Trade Center was the result of a controlled demolition rather than structural failure due to impact and fire and that the Pentagon was hit by a missile launched by elements from inside the US government in order to justify the invasions of Afghanistan and Iraq as well as geostrategic interests in the Middle East, such as pipeline plans launched in the early 1990s by Unocal and other oil companies?

**QUESTION BART SIMPSON:** Are swim skirts cute or what ;)? LOVE YOU BABES SO MUCH!

# WEIGHT-LOSS UPDATE

My baby-gals!!! We're between chapters! I am so happy you're here. Let's just say I've been going through a rough time. It's been another seven months since I last wrote. I had to really do a lot of soul-searching as a very fat, single person. Did you know that your soul also gains weight when you become a fat person? My soul is definitely one of my problem areas. For me, my soul was obese.

But, after months and months and yo-yoing like crazy, I finally stopped *searching* for myself and *found* myself. After that, the pounds just melted off. My weight-loss regimen has been extremely successful. I've lost the cake weight and then some!! **FIG. 3.8** How bomb-ass do I look, ladies!!! What's my secret, you ask (can I get a "What's your secret?"!)? Oh, just the regular old stuff—diet, exercise, and meth. Yes, you read that right! Methamphetamines! You know: Crystal! Crank! Street-meat! Smashmouth! TaTu! Poopoo doodoo! Don't Meth with Texath!

You wouldn't think it, but meth is actually an extremely easy way to lose weight *and* stay up for nine days at a time. You lose like sixty pounds in teeth alone. I don't know why I didn't think of it before—I've been able to quadruple

my writing time and negative-quadruple my weight. And I feel great on it. I don't even have time to think about Xander because I'm too busy BJing my dealers so they'll throw in an extra bump because I need it. Find me someone who spuriously asserts that meth turns you into a psychotic, aggressive monster and I will rip that person's head off until all of the blood comes out. Just call me "Methin' Amram!"

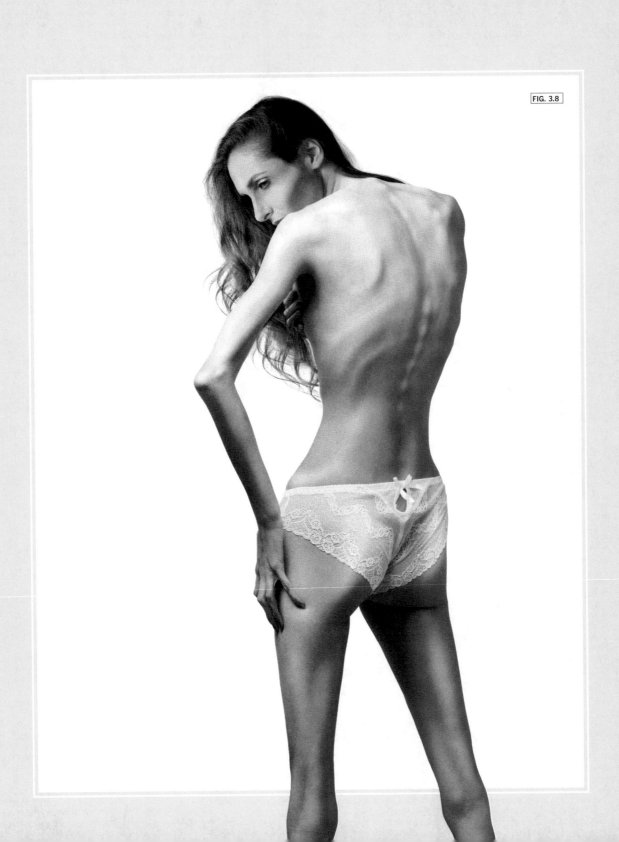

FIG. 3.8

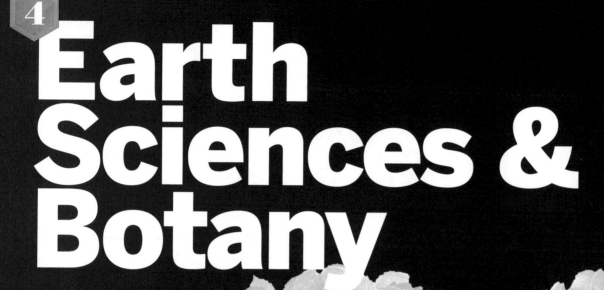

# 4

# Earth Sciences & Botany

# Introduction

"Earth sciences & botany" comes from the English *earth sciences & botany*, which, loosely translated, means "earth sciences & botany."

Botany specifically studies the science of plant life, while earth sciences looks at the functioning of our planet, drawing from other scientific disciplines to try to understand how the planet Earth works and how it evolved to its current state. Like, back in the day, Earth used to be a molten pile of dirty magma (SEX MOVE DU CHAPTOUR, MADEMOISELLES!! "Molten Pile of Dirty Magma" is where you take a "molten pile" on your partner's chest).

This is the section where we delve into the science of Mother Earth. Ladies, we all have mothers! It's fun to think of Mother Earth like your own mother! FIG. 4.1

FIG. 4.1

HOW
MOTHER EARTH
IS LIKE
MY MOTHER

 =

Going a little bald up top! (hole in the ozone layer)

Hot flashes! (global warming)

Big fatso! (shaped like a perfect sphere)

Disgusting tramp stamp! (Mozambique)

Complicated relationship with her daughter! (earthquakes)

Our understanding of the earth has changed greatly over the years. Today's scientists didn't know anything years ago! Probably because those scientists were babies. Other scientists, who were older, may have known some things about the earth.

Like, here's a crazy fact for you: people used to think that the earth was flat. Just like my friend LAUREN, HAW HAW HAW!

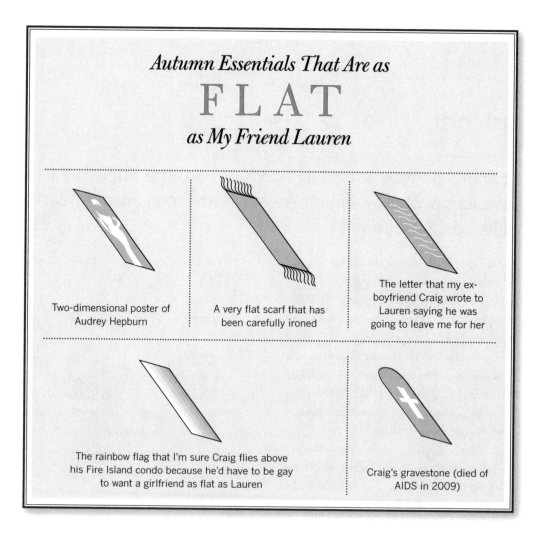

## Autumn Essentials That Are as
# FLAT
## as My Friend Lauren

Two-dimensional poster of Audrey Hepburn

A very flat scarf that has been carefully ironed

The letter that my ex-boyfriend Craig wrote to Lauren saying he was going to leave me for her

The rainbow flag that I'm sure Craig flies above his Fire Island condo because he'd have to be gay to want a girlfriend as flat as Lauren

Craig's gravestone (died of AIDS in 2009)

# Four-Chapter Anniversary

Before we go any further, let's take a moment to celebrate the FOUR-CHAPTER ANNIVERSARY of *Science . . . for Her!* We've been through so much together! Let's take a look back at some of *Science . . . for Her!*'s best pieces from the past four chapters!

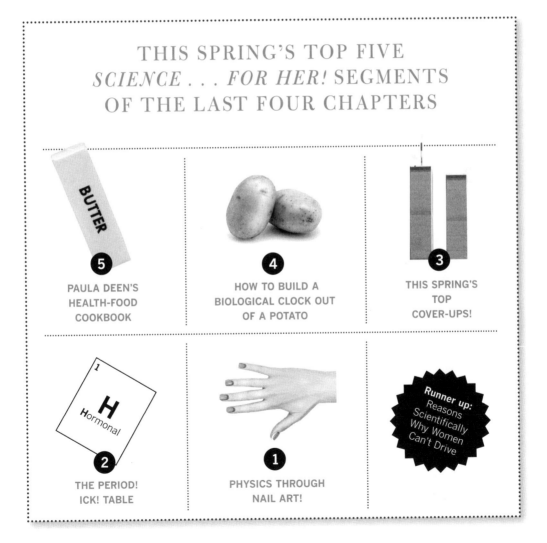

THIS SPRING'S TOP FIVE
*SCIENCE . . . FOR HER!* SEGMENTS
OF THE LAST FOUR CHAPTERS

**5** PAULA DEEN'S HEALTH-FOOD COOKBOOK

**4** HOW TO BUILD A BIOLOGICAL CLOCK OUT OF A POTATO

**3** THIS SPRING'S TOP COVER-UPS!

**2** THE PERIOD! ICK! TABLE

**1** PHYSICS THROUGH NAIL ART!

**Runner up:** Reasons Scientifically Why Women Can't Drive

CHECK IT OUT!

# Drugs Not Hugs

Let me take a slight little digression before we get into the meat and potatoes (more like meat and extra meat—potatoes have carbs, girls!) of earth sciences and botany. This seems like as good a chapter as any to talk about drugs, since many of them are plants or are made from plants from the earth or made from the hand of God (meth).

Some people say "Hugs, not drugs." I always say drugs *and* hugs! In fact, drugs often make hugs feel way better. Go take a bunch of Ecstasy and hug a shag carpet. You'll thank me later. But I'll focus for now on more plant-based drugs. *Marijuana* or *weed* is a fun gateway drug! You can get prescriptions for marijuana to help treat ailments such as glaucoma or addiction to marijuana. FIG. 4.2

*Rohypnol* or *roofies* is another fun gateway drug . . . gateway to *dating!!!!!* FIG. 4.3 Girl, haven't you dreamt that someday you'll find the perfect guy who will sweep you off your feet because you

FIG. 4.3

FIG. 4.2

**WEED STRAIN**
~~~ OR ~~~
*Celebrity Baby Name?*

- OR -

| 1 | 2 |
|---|---|
| Casey Jones | Blue Dream |
| 3 | 4 |
| Hawaii Gold | Purple Haze |
| 5 | 6 |
| NY Diesel | Pure Power |
| 7 | 8 |
| Sky Walker | Charlie Sheen |

WEED: 1, 2, 3, 4, 5, 6, 7 BOTH: 8

can't stand on your feet because you've dropped unconscious due to the roofies that he's put in your mai tai?!?! You are not alone!!! Every gal wants to wake up in a handsome stranger's bedroom/secret basement cage! Also, roofies have *no* carbs! They are pretty much a perfect drug.

And, as I mentioned previously, meth is truly a drug of the gods. Like the little saying goes, "Candy is dandy but meth is bestth." After becoming violently addicted to meth, I've been awake for nearly a fortnight and have eaten only three raisins and a stack of ants that dared try to challenge me to a dance-off. Of course I'm going to fucking beat you, ants, I'm an Egyptian god.

# Dendrology: The Study of Trees

Okay, this is real classic botany stuff, right here. Trees, man! Growing up, you were probably told that trees were a wonderful part of being alive, and that we should do everything in our power to conserve them. WRONG. Trees are every woman's NATURAL ENEMY, YOU DUMB BITCH.

The tree lobby (Big Tree) has managed to convince our generation that trees are great and give you air and stuff (verbatim quote from Big Tree). But how about the fact that approximately 80 percent of them will become copies of *Mein Kampf*?! I personally like to flip off trees just in case they turn into Guy Fieri cookbooks. But most important: they give men completely unrealistic ideals for women's bodies. We can't all be effortlessly tall and thin with brittle skin!! That's why I've included five blank pages in *Science . . . for Her!* If I can kill even *one* more tree than usual with the publication of this book, then I'm doing my part for womankind.

Doesn't that feel amazing!!! Take *that*, you big stupid trees with dumb leaves (it's like, who has *leaves* anymore?! You untrendy bitch) and idiot roots (hey dummy, your roots are showing)! FUCK YOU, TREES!

Really the only thing a tree is good for is when you KILL IT (long and painful death, please!) and make it into fun and/or flirty books like this one or other cute ones. Here are some of the books from Urban Outfitters that I'm most excited to impulse-buy next season! `FIG. 4.4`

If we're talking about trees, we HAVE to talk about the *Christmas tree*. As a Jew, I grew up thinking it was the worst kind of tree of them all, but I have to say, I did come around to Christmas after I celebrated it with my then-true-Christian-love Xander. Hey, girlies, why don't I tell you of my FIRST CHRISTMAS! This should kill a few ol' *Tannenbaums*!

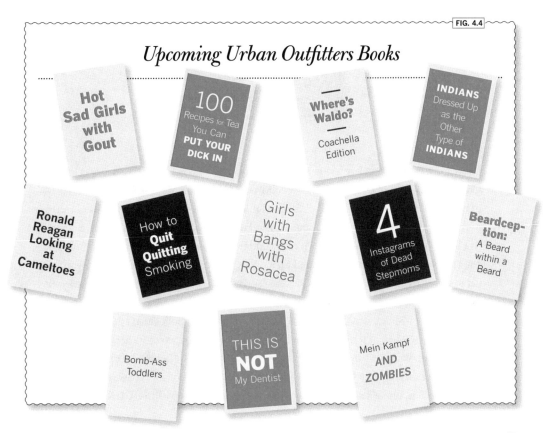

`FIG. 4.4`

## *Upcoming Urban Outfitters Books*

Hot Sad Girls with Gout

100 Recipes for Tea You Can PUT YOUR DICK IN

Where's Waldo? Coachella Edition

INDIANS Dressed Up as the Other Type of INDIANS

Ronald Reagan Looking at Cameltoes

How to Quit Quitting Smoking

Girls with Bangs with Rosacea

4 Instagrams of Dead Stepmoms

Beardception: A Beard within a Beard

Bomb-Ass Toddlers

THIS IS NOT My Dentist

Mein Kampf AND ZOMBIES

# MY FIRST
## *Christmas*

Because I was raised as a Jew, a woman, and a Democrat, I was denied the luxury of celebrating Christmas. In December, when the other children would dream of sugarplums, I would dream of regular plums. When the other children would dream about nutcrackers, I would dream of regular plums (I'm super into plums). I just knew Christmas as that day when my family would eat Chinese food and watch television and I'd sneak some extra plums from Nana's old-country plum basket. Sure, I celebrated Hanukkah, but who wants a holiday where you don't get to have a tree or whine about dogs or eat lasagna? Not Garfield the Cat, let me tell you!

That first year with Xander, though, I learned that Jews and Catholics aren't that different! First of all, we both believe in Jesus as our One Savior (if you round up to the nearest whole Savior)! Second of all, we both hate Mondays but love lasagna! I might be mixing up Catholics and Garfields for that one, but you get the point—I was ready to delve into a new world of cultural understandings, capital letters, and ® symbols. I was ready for my FIRST CHRISTMAS®®®®.

We began my Christmas journey® (not to be confused with the band Journey®, although it was rumored that in the eighties, on the road, they experimented with Christmas) the day after Thanksgiving, when Xander bought our fake Christmas tree at Bed Bath & Beyond. Before someone makes a joke about the tree probably being from the "Beyond" section, I have to warn you—the tree is also a bath! So it's from the "Bath" section! There's a hollow in the center that you can fill with hot water and carefully bathe one leg in at a time. It's a fake tree, obviously, but I would still throw it away and buy a new one every year to get the full Christmas effect® (not to be confused with *The Butterfly Effect*® starring Ashton Kutcher®, a Jew®).

Christmas ornaments are the part of Christmas that I find most confusing. Who needs Christmas ornaments when nature has provided her own ornaments: bird's nests, Frisbees, and plums that I sometimes Scotch-tape onto non-plum trees for fun? Still, decorating the tree was pretty cool. Xander provided the colored orbs. I provided the ol' tape-plums. Garfield provided the laughs.

As I then threw our stockings on the blazing fire to heat us in that cruel Hollywood winter, Xander explained to me the story of Santa Claus. I realized the name "Santa Claus" is a play on the movie *The Santa Clause* starring Tim Allen. Super cute reference, Catholics!! Slowly, the unfamiliar sounds of Christmas music filled our two-bed, two-bath, four-kitchen, fifty-two-door apartment. I instantly loved it! Especially the part in "Winter Wonderland" where they talk about pretending a snowman is Parson Brown. A marriage is a sacred institution between one man and one woman married by one ordained snowman, and this song does justice to my staunch Catholic principles.

Though I'd tackled the big Christmas to-dos, I still had a lot to learn. The true meaning of Christmas cannot be bought with a Papa John's Pizza gift card (they sell lasagna, too!) or looked up in a dictionary under the definition of "Christmas"; it must be discovered through careful observation of true believers. So, until December 25 every year, I'll be watching you, Christmas lovers. Like a marathon of all the episodes of the *Garfield & Friends*® animated TV show that ran from 1988 to 1995, I'll be watching you®.

# Floriculture: The Study of Flowers

Oh. My. GAAAAAAH. TUH. I am seriously sobbing, I am so excited to start talking about flowers!!! If I could've put flowers into every other single part of science, you bet I would've. I should've called atoms "microscopic flowers." I should've called periods "panty flowers."

A *flower*, sometimes known as a *blossom* or a *gay tree*, is actually the reproductive structure of certain plants. So when you're joking around with your best friends that certain flowers look like vaginas, you're not too far off! `FIG. 4.5` Actually, speaking of best friends, one of my best friends, Minnie, is literally a flower. She's a baby's breath! She's so beautiful, and never asks me to drive her to the airport!

`FIG. 4.5`

Flowers are really great for day-to-day life. Flowers make for great tattoos on the small of your back. How amazing is that! You can put a cute plant's vagina on your back! The names of flowers can be used to name sluts and other types of ladies. Have you ever met a Violet or a Jasmine who wasn't giving a for-profit blowjob while she was shaking your hand to meet you? Best of all, flowers are a great thing to put onto *fabrics*, which, as you may know, is often the main component of *clothes*, also known as *gay skin*.

## *Sexy or Skanky?*
### FLOWER EDITION

**There is a very fine line between sexy and skanky in the flower world. Don't catch yourself stepping over the line!**

**ROSE:** SEXY
So classic, so elegant, so fun, so flirty! The rose has rarely if ever made a fashion misstep.

**CROCUS:** SKANKY
C'mon, pull yourselves together, crocuses! You're just asking for it by dressing like that!!

**DAISY:** SEXY
Sunny like a summer's day with just a little innocent petal showing! Daisies truly understand that it's what you *don't* show that's sexy.

**ORCHID:** SKANKY
Pull something on over that long stem, you big slut! Orchids are basically vaginas on a fishing rod! Eww!

**LILY:** SEXY
Perfect symbol of sweetness. A lily on the streets and an orchid in the sheets!

**LILY OF THE VALLEY:** SKANKY
See, this is what I'm telling you. VERY fine line. Lilies = sexy. Lilies of the VALLEY = skanky as heck. What kind of a *lady* hangs out in a *valley*?! Also, this flower isn't even eighteen!! Where are her parents?!

Flowers are an amazing form of human confabulation as well. Holy crap, ladies—I finally used *confabulation* in a sentence! I have a word-a-day calendar and that was today's word! Well, not today. That was a word from last June, but I haven't been able to use it in a sentence until today. It's pretty rough because I haven't known what day it was since June 14, which is the day that I first got *confabulation*. I get to finally turn the page!! Let's see . . . June 15 is . . . *salivate*. Fuck.

Yes, so flowers are a great way of *confabulating*, or communicating. A bouquet from a boy can say as many varied things as "Happy birthday," "Let's do it backwards tonight," and "I stole this bouquet from a Ralph's, see, the tags are still on." FIG. 4.6 If you're wearing a sundress that has a flower on it, a boy usually knows that you're nasty as f.

FIG. 4.6

So . . . Backwards then?

"Speaking" of communication (fpun [fun-pun] intended), I just can't say enough good things about flowers. They're admirable, alluring, angelic, appealing, beauteous, bewitching, charming, classy, comely, cute, dazzling, delicate, delightful, divine, elegant, enticing, excellent, exquisite, fair, fascinating, fine, foxy, good-looking, gorgeous, graceful, grand, handsome, ideal, lovely, magnificent, marvelous, nice, pleasing, pretty, pulchritudinous, radiant, ravishing, refined, resplendent, shapely, sightly, splendid, statuesque, stunning, sublime, superb, symmetrical, taking, well formed, and wonderful. I guess what I'm trying to say is, I am just trying to use as many words as I, Megan Amram, possibly know so that I, Megan Amram, the author of this book, FIG. 4.7 can kill (i.e., murder, assassinate) as many trees as possible.

FIG. 4.7

# Geology: The Study of Solid Earth

Geology is the science of solid earth, the *rocks* of which it is composed, and how they change. Rocks remind me of one of my favorite song lyrics: "My name is Kiiiiiiid Rock!!!" (I forget who sang it???)

Rocks aren't just gross dirt clods—they're also gemstones! I'm talking *DIAMONDS*! You might know diamonds from the anonymous quote: *"SHINE BRIGHT LIKE A DIAMOND! My name is Rihanna!"* Diamonds are formed when the carbon in *coal* is compressed and heated under very high pressure and temperature. All diamonds begin as coal. Neil Diamond was named Neil Coal until he was like thirty-seven.

You may have heard about *conflict diamonds*, also known as *blood diamonds*. These refer to diamonds mined in war zones and sold to finance insurgencies, in places like *Sierra Leone* and *Liberia* and other *African countries*. Sorry to be TMI, but: I love them!!! Not African countries, silly—blood diamonds! Honestly, I find that blood diamonds often have a much prettier sheen than non-blood diamonds. Something about the guns that are going off murdering people around them seems to make the diamonds sparklier. Maybe I'm imagining it, but I don't think so! Like I've always said, BLOOD diamonds are a girl's BLOODST FRIEND!

> **Good Name for Tiffany's Vag I Just Thought of**
> "Blood Diamond"

---

*Why Diamonds Are*
BETTER BEST FRIENDS
*Than My Friend Tiffany*
(no offense)

**Many people have said, "Diamonds are a girl's best friend!" Here are just a few reasons that diamonds are so much better best friends than my slutty ex–best friend Tiffany!**

Diamonds are less ugly than Tiffany

⬦⬦⬦

Diamonds won't ever f your ex Xander

⬦⬦⬦

Diamonds go with everything; Tiffany only "goes with" (sleeps with) pieces of trash

⬦⬦⬦

Tiffany, you smell like a barfed-up bagel if it had a million stretch marks

⬦⬦⬦

Also your vag is always beet red no matter the day. Xander told me that in confidence and now I am embarrassing you with it

⬦⬦⬦

JK love you, babe

⬦⬦⬦

JK

# Global Warming:
# The Study of Global Warming

The sun is shining, the birds are singing, you know what that means—it's February! Thank you, GLOBAL WARMING!!

*Global warming*, more accurately known as *climate change*, is the phenomenon of shifting weather patterns. Contrast with *gay-ball warming*, which is when a gay person's balls are cupped by another gay person. When people talk about climate change, they're usually talking about the rising global temperature and pollution that has arisen from man-made change. Make that *woman-*made, sister! Ruining the earth's climate isn't just for men anymore! You *too* can eff this planet up!

Over years of industrialization, humans have greatly influenced the makeup of the earth's atmosphere. Fossil-fuel consumption, aerosols, and deforestation have all played their parts in global warming. But, like, are you seriously asking me to give up all my favorite things just so the earth doesn't get hotter? Like, my favorite things are *literally* driving around in pink Hello Kitty Hummers, hairspray, and deforestation. I literally just wrote a part of this book about killing trees. In fact, let's kill just a little more of a forest.

Some women (that's US, GIRLS!) don't believe that global warming is real. They think it's a cute little myth, like Santa Claus/the Tooth Fairy/female orgasms. I, on the other hand, know that it's real and that I *love* it! Who doesn't like warm weather all year round, fewer bees, no forests? The earth is literally hotter than Taylor Lautner, Robert Pattinson, and Rihanna (that's for you lesbos!) combined! It's so hot in LA right now that my ex-boyfriend's house that I just set on fire is on fire!

I think the reason some gals don't want to face the climate-change music is that it's a little scary. It's easy to not face things that are bad in your life. No one wants to *salivate* at that (NAILED IT, FUCK YOU WORD-A-DAY CALENDAR! Let's see what's next . . . *perspicacious*?! You gotta be fucking kidding me!). Like, if I'm being honest, my relationship with Xander was doomed fairly early on. I should've known we were heading for a breakup. Xander would give me these little hints like saying to my face "we are breaking up" and getting a restraining order and then saying "we are broken up," etc. etc. I was just like, why don't you just come out and *say it*, coward?!

I should probably be sad about global warming, but I'm not. I'm ready to tackle it like the modern woman that I am. I don't know if it's the beautiful weather outside or the meth I just smoked, but I feel as giddy as someone who's just smoked meth!!!

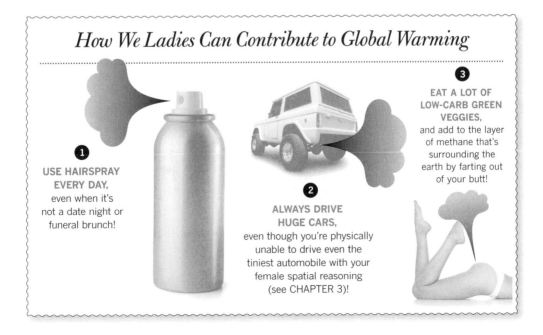

## *How We Ladies Can Contribute to Global Warming*

**1**

**USE HAIRSPRAY EVERY DAY,**
even when it's not a date night or funeral brunch!

**2**

**ALWAYS DRIVE HUGE CARS,**
even though you're physically unable to drive even the tiniest automobile with your female spatial reasoning (see CHAPTER 3)!

**3**

**EAT A LOT OF LOW-CARB GREEN VEGGIES,**
and add to the layer of methane that's surrounding the earth by farting out of your butt!

# FASHION STAPLES *for*
# EACH STEP OF GLOBAL WARMING

**As the world becomes progressively warmer, the modern woman's wardrobe is going to have to change accordingly. Here are fashion options and transitions for each 10-degree Fahrenheit increase up until the point of complete human destruction.**

## +10°
### FAHRENHEIT

At a 10-degree global temperature increase, the world as we know it has completely transformed. This seemingly insignificant change in temperature has led to widespread heatstroke, natural disasters, and fatalities. A cute **spaghetti-strap tank** is perfect. They come in every color, so let your unique personality shine through!

## +20°
### FAHRENHEIT

Most of Africa and Australia is now unlivable for humans. Half of the human population has been wiped out due to lethal heat-stress levels. Try a super-light **linen sundress** in white or light yellow for a summery splash of color.

## +30°
### FAHRENHEIT

The entire planet is now unlivable. Ice caps have all melted, creating planet-wide oceans that cover every land mass. Since you'll be underwater, make sure you have the perfect swimwear. **Tankinis** are *in*! Just make sure your beach body is in, too! ;)

## +40°
### FAHRENHEIT

The planet formerly known as Earth is unrecognizable. There is no land or water—Earth has become a gaseous mass like Jupiter. Humans and any evidence of human life have long since perished from the planet. And since there are no people around to judge your outfits—don't worry about your clothes! Go out in your **yoga pants**! How fun is that!

# The O-Zone:
# The Study of the Female Orgasm

Speaking of ozone layers . . . we're talkin' female orgasms in a section I like to call the **O-ZONE!!** Fpun fun-tended!

Like I promised in the biology chapter, let's talk more about how female orgasms don't exist. Any girl who is telling you that she's ever had an orgasm is lying to your face. She's probably just trying to make you feel bad (some best friend *she* is!) or trying to make her boyfriend seem adequate. If you've ever had what you think is a female orgasm, you probably actually just ate a good grape.

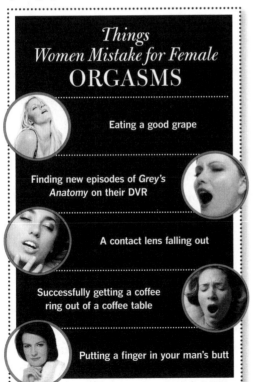

*Things*
*Women Mistake for Female*
ORGASMS

Eating a good grape

Finding new episodes of *Grey's Anatomy* on their DVR

A contact lens falling out

Successfully getting a coffee ring out of a coffee table

Putting a finger in your man's butt

# The Apocalypse

I was about to write about the *apocalypse*, but it appears the apocalypse has come EARLY FOR ME BECAUSE I FREAKING RAN OUT OF METH, GALS. You never think it's going to happen to you, but then it does and no one's there to help you. I'm so stressed about it, I've been grinding my two teeth all night. FIG. 4.8

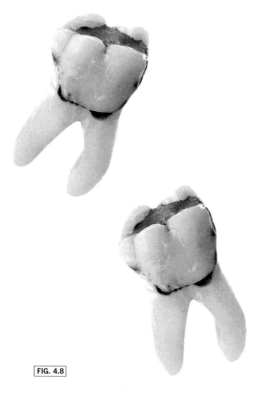

FIG. 4.8

Some scientists predict that the world is going to end in 2063. You know what that means, ladies—as of the writing of this, only 17,801 shopping days until the end of the world!! Make sure to get to the malls early. They are going to be *packed* the day before the apocalypse. Lots of "everything must go!" sales.

No matter how great we live our lives, the end of the world is coming at some point. That's why you have to live it up while you can! Learn a new hobby. Kiss your crush. Kill Xander's cat. FIG. 4.9

FIG. 4.9

Skip forward a few days in your word-a-day calendar. Punch a bat in the face. Buy a dress made of *gossamer*, the word of the day for July 5. Eat a *dick*, the word of the day for July 7.

I'm pleased to present a *Science . . . for Her!* SNEAK PEEK of the NEWEST *Sex and the City* movie, which takes place in the postapocalyptic wasteland that was previously America! We got our hands on a scene from when Carrie and your other fave girls have to juggle dating, work, and being alive in the smoldering ashes of the apocalypse!

# Sex & the Postapocalyptic Dystopian Landscape

(CARRIE, MIRANDA, CHARLOTTE, and SAMANTHA sit around a brunch table in a smoldering crater after the apocalypse. They sip mimosa glasses filled with scarab beetles and blood. SAMANTHA is on fire.)

CARRIE:
Ladies, have you ever noticed how hard it is to find a good man in this postapocalyptic dystopian landscape?!

MIRANDA:
Have I ever!! It seems like all the good men are either taken or don't have skin because their skin melted off during the apocalypse because of the acid rain that melted off their skin!

CHARLOTTE:
Or are Jewish!

(They laugh.)

MIRANDA:
And it's near impossible to find an affordable apartment in Manhattan. Or should I say, IneedaMAN-hattan. Or should I say, the ocean. Manhattan is now under the ocean.

CHARLOTTE:
Or are Arabs!

CARRIE:
Ladies, a toast. (CARRIE, MIRANDA, and CHARLOTTE raise their glasses. SAMANTHA raises her severed, melting left arm in her right arm.) We'll be together forever! Or until the next lava-bee swarm attacks, which is in approximately one one-hundred-hundred hundredth of a nanosecond.

(A lava-bee volcano-stings SAMANTHA in the mouth. She's slutty about it.)

CHARLOTTE:
Hey, do you know if those bees are single? I would date a bumble.

CARRIE:
I would FUCK THAT BUMBLE.

CHARLOTTE:
Or are short Mexicans!

SAMANTHA:
Bdddbdbdbddddbddd.

(Her skin melts off.)

CHARLOTTE:
Oh, Samantha, you SURE are the slut of the group!

(They drink to SAMANTHA'S sluttiness. Her lips fall off into her beetle-mosa.)

CHARLOTTE:
So, ladies, I went out on a date on Wednesday night, and it was TERRIBLE. He was just a burnt-up THORAX. Pro: he comes from a good family and is a good listener. Con: he's a THORAX.

CARRIE:
How was the sex?? DISH!

CHARLOTTE:
I put my boob on his thorax.

MIRANDA:
NOW THAT'S THE KIND OF RACY GIRL TALK I'M TALKIN' ABOUT, GIRLS! WHO CARES IF IT'S THE END OF THE WORLD, WE'RE DISHING DISHING DISHING ABOUT MEN!

CHARLOTTE:
Or are short or tall Gypsies or Africans!

(They drink to dishing about men. SAMANTHA'S eyes fall out into her beetle-mosa.)

CARRIE:
HEY, the only man I need is a pair of Manolo Blahniks. And some potable water. Some extra blood. Religion. A hand to replace my hand, which is about to fall off. (Her hand falls off.)

MIRANDA:
But seriously, how are men not flocking to us? We're smart, sexy, and have even more sexable holes in our bodies than before the apocalypse! I would let a man boom-boom the hole in my calf that a meteorite went clear through this morning.

CHARLOTTE:
I would boom-boom some salt.

CARRIE:
I would boom-boom a sick child's coloring book.

MIRANDA:
I would boom-boom a sexy leek.

CHARLOTTE:
I would boom-boom some sexy sand.

MIRANDA:
I would boom-boom a sexy ransom note.

CHARLOTTE:
I would boom-boom some hope.

CARRIE:
I would boom-boom a sexy hurt horse.

SAMANTHA:
Bd. Db.

(Her head falls off. One tear comes out of a smoldering hole in her upper arm. That's where she cries from now.)

CARRIE:
You don't look so good, Samantha.

MIRANDA:
Look who's talking, Sarah Jessica Parker.

There are many ways that the apocalypse might occur. Famous authors have loved to try to predict how mankind (or WOMANKIND!) will react to the end of human civilization, due to war or global warming or a combination thereof. For example, the children's book *Goodnight Moon* is basically like a fun telling of the end of everything in known existence including the moon. FIG. 4.10 Ayn Rand, a woman (can I get a *gossamer*, the word of the day for July 5?!), wrote a book called *Anthem* about a dystopian future. What a crazy bitch!

Because we at *Science . . . for Her!* think that women should help other women, we've enlisted that crazy bitch Ayn Rand to answer some of your most pressing questions! XOXO love you, hope this helps!

FIG. 4.10

—Ahh!

# Dear Ayn Randers

*Dear Ayn,*
*I'm dating a man who I think I love, but I'm afraid he's having an affair. He comes home late, he acts suspiciously, and he even has red lipstick on his collar. Should I confront him or just hope for the best?*
*County Af-fair*

Dear County,
Red lipstick? Your husband is a Communist. Divorce him and sell his clothes, children, and pens to make money to spend on cars, human slaves, and bigger pens. This will simultaneously stimulate the economy and punish the slaves for not having jobs. Slaves: what lazybones!
Hope this helps,
Ayn

―――――――――

*Dear Ayn,*
*I'm trying to figure out which color dress to get my daughter for her First Communion. Is red gauche?*
*Paint the Dress Red*

Dear Paint,
Hmmm, this is a tough one. On the one hand, I hate Communism ("Reds"). On the other hand, I hate

religion. On the third hand, I hate women. FYI, do you know how I got that third hand? I bought it from a child! Ho HO! He was easily tricked into selling me his hand for a nickel and a pious man's drum! I have a baker's dozen child-hands in my glove compartment!!!!
Hope this helps,
Ayn

―――――――――

*Dear Ayn,*
*My baby daughter is turning one year old, and I don't know if I should throw her a birthday party or not. What should I do? I'd appreciate any advice.*
*One Is the Loneliest Number*

Dear One,
DO NOT reward this tiny unemployed Jew with a party. Your so-called baby is most likely an immigrant (read: LAZYBONE) who doesn't contribute to her family's income and gives terrible, poor-people gifts like HD-DVDs and sand. Unrelated question: does your baby have any spare hands?
Hope this helps,
Ayn

*Dear Ayn,*
*My in-laws are coming to Thanksgiving dinner at my house for the first time. I'm not great at hosting: how do I make sure we have enough food and that we all get along?*
*Turkey Lurkey*

Dear Lurkey,
The lavish Thanksgiving meal is a symbol of the fact that abundant consumption is the RESULT AND REWARD OF PRODUCTION. Do you see a poor "person" on the street? (NOTE: I put "person" in quotes because poor people are more like CHAIRS in my book because you should SIT on them.) Ask this "person" (read: chair) for his half sandwich for your Thanksgiving meal. Does he not relinquish that symbol for all American pride, the half sandwich? Does he not relinquish his half BLT, his half PB&J? Distract him with some sort of juvenile puppet-based theater and steal that half sandwich. That is YOUR HALF SANDY, for YOU ARE GOD. YOU ARE GOD. YOU ARE A GOD EATING A HALF HAM-AND-CHEESE SANDY. Note: to be clear, it is half of a ham-and-cheese sandwich,

not a whole half-ham-and-cheese sandwich.
Hope this helps,
Ayn

---

Dear Ayn,
*Are you the warrant and the sanction?*
*Dawdling in Dallas*

Dear Dawdling,
I am the warrant and the sanction.
Ayn

---

Dear Ayn,
*I don't mean to be offensive, but your writing is overwhelmingly juvenile and one-note. How did you become such an influential figure, a cornerstone of the landscape of American conservative politics? You write like a petulant child.*
*Holly Hurlbut*
*Professor of Comparative Literature*
*Harvard University*

Dear Holly,
Your mom's juvenile.
Hope this helps,
Ayn

Dear Ayn,
*I'm in Los Angeles for a day and I don't have much spending cash. What are some fun things to do that are cheap and easy?*
*SoCal SoCheap*

Dear SoCal,
Here are some options:
- Tattoo "laissez-faire" on a celebrity's bagel.

- Build a statue of me, Ayn Rand, out of cheap materials (rose gold, the word of a liberal, Mexican day labor).

- Throw that statue at the chair who built it (aim for the throat).

- Go to the zoo and taunt an animal smaller than you (human children count).

- Make a coat out of some Dalmatians.

- Push a baby into another baby and point and laugh while they cry and then trip the babies and then laugh more at those babies that you tripped.

- Make a coat out of someone with Medicare.
Hope this helps,
Ayn

Dear Ayn,
*I'm considering becoming a Communist. Should I become a Communist?*
*Commie Dearest*

Dear Commie,
No.
Hope this helps,
Ayn

---

Dear Ayn,
*If I yell enough at gays and Jews and Mexicans and Michael J. Foxes, will my daddy love me? Will he kiss me on the face and not throw paperweights at my face and love me?*
*Rush Limbaugh*

Dear Rush,
Yes.
Hope this helps,
Ayn

---

Dear Ayn,
*I caught my wife reading* Atlas Shrugged *the other day. She's been acting strange ever since: yelling for no reason, physically harming children, stealing from those poorer than us, hating other women. Do you know what's wrong?*
*Sincerely,*
*Atlas My Love Has Come Again*

Dear Atlas,
Women can't read.
Ayn

# Earth Sciences & Botany Recap

I hope you had as much fun as I did with this chapter. Now that we're four chapters into this book, I understand if you're getting a little tired. All that I can say is, learning science is a marathon, not a sprint. Make sure not to shirk your other feminine duties just so you can read this book. Take care of your family. Pleasure your husband. There are some things more important than science, if you can believe it!

I literally had a blast writing this chapter for you, bebs. It's incredible how quickly you can write things when you don't sleep. Also, it's incredible how many times you piss yourself if you don't pause to go to the bathroom when you're awake for a million hours. This alleyway that I've been living in for a few weeks is *filled* with human urine!

# Recap Questions

**QUESTION 1:** Which *Sex and the City* girl is an orchid most like? (HINT: ORCHIDS ARE SLUTS & BRATZ!)

**QUESTION 2:** Can I stop whenever I want?

**QUESTION 3:** Of these things, what does my friend Lauren most resemble: flatironed hair, piece of flat paper, or double-mastectomy postop woman?

**QUESTION 4:** Do I hate trees? NOTE: I've included a full blank page for you to write your answer.

**QUESTION 5:** Can I feel even closer to you than I do right now? I feel so close to you right now it's insane. LOVE YOU SO MUCH I WANNA KILL MYSELF!

# Chapter 5

Xan-
Xander,     bo-bander
Fee-Fi-mo-mander, Xander! Xan-
fana fo fander Fee-Fi-mo-mander, Xan-
fana fo fander Fee-Fi-mo-mander, Xander! Xan-
Fee-Fi-mo-mander, Xander! Xander, Xander, bo-
Xander! Xander, Xander, bo-bander Banana-fana fo
bo-bander Banana-fana fo fander Fee-Fi-mo-mander, Xan-
Fee-Fi-mo-mander, Xander! Xander, Xander, bo-bander Banana-fana fo fander
Xander, bo-bander Banana-fana fo fander Fee-Fi-mo-mander, Xander! Xander, Xander, bo-bander Banana-fana fo fander Fee-Fi-mo-mander, Xander! Xander, Xan-
der, bo-bander Banana-fana fo fander Xander, bo-bander Banana-fana fo fander Fee-Fi-mo-mander, Xander! Xander, Xander,
Fi-mo-mander, Xander! Xander, Xander, bo-bander Banana-fana fo fander Fee-Fi-mo-mander, Xander! Xander, bo-bander Banana-fana fo fander Fee-
bo-bander Banana-fana fo fander Fee-Fi-mo-mander, Xander! Xander, Xander, bo-bander Banana-fana fo fander Fee-Fi-
Fi-mo-mander, Xander! Xander, Xander, bo-bander Banana-fana fo fander Fee-Fi-mo-mander, Xander! Xander, bo-bander Banana-
Xander, bo-bander Banana-fana fo fander Fee-Fi-mo-mander, Xander! Xander, Xander, bo-bander Banana-fana fo fander Fee-Fi-
fana fo fander Fee-Fi-mo-mander, Xander! Xander, Xander, bo-bander Banana-fana fo fander Fee-Fi-mo-mander, Xan-
mo-mander, Xander! Xander, Xander, bo-bander Banana-fana fo fander Fee-Fi-mo-mander, Xander! Xander, Xander,
der! Xander, Xander, bo-bander Banana-fana fo fander Fee-Fi-mo-mander, Xander! Xander, Xander, bo-
Xander, bo-bander Banana-fana fo fander Fee-Fi-mo-mander, Xander! Xander, Xander,
bo-bander Banana-fana fo fander Fee-Fi-mo-mander, Xander! Xander,
bander Banana-fana fo fander Fee-Fi-mo-mander,
Xander, bo-bander Banana-fana fo fander
Xander! Xander, Xander, bo-bander Banana-fana fo fander
Fee-Fi-mo-mander, Xander! Xander, Xander, bo-
bander Banana-fana fo fander Fee-Fi-mo-mander, Xander! Xander, Xander, bo-bander
Banana-fana fo fander Fee-Fi-mo-
mander,  Xander!  Xander,
Xander, bo-bander Ba-
nana-fana fo fander
Fee-Fi-mo-
man-

d e r ,
Banana-fana fo fander
der, Xander, bo-bander Banana-
der! Xander, Xander, bo-bander Banana-
der, Xander, bo-bander Banana-fana fo fander
bander Banana-fana fo fander Fee-Fi-mo-mander,
fander Fee-Fi-mo-mander, Xander! Xander, Xander,
der! Xander, Xander, bo-bander Banana-fana fo fander
Banana-fana fo fander Fee-Fi-mo-mander, Xander! Xander,
Xander! Xander, Xander, bo-bander Banana-fana fo fander Fee-
fander Fee-Fi-mo-mander, Xander! Xander, Xander, bo-bander Banana-fana fo fander Fee-Fi-mo-mander, Xander! Xander, Xan-
der! Xander, Xander, bo-bander Banana-fana fo fander Fee-Fi-mo-mander, Xander! Xander, Xander,
bander Banana-fana fo fander Fee-Fi-mo-mander, Xander! Xander, bo-bander Banana-fana fo fander Fee-
fander Fee-Fi-mo-mander, Xander! Xander, Xander, bo-bander Banana-fana fo fander Fee-Fi-
mo-mander, Xander! Xander, Xander, bo-bander Banana-fana fo fander Fee-Fi-
Banana-fana fo fander Fee-Fi-mo-mander, Xander! Xan-
der! Xander, Xander, bo-bander Banana-fana fo fander Fee-Fi-mo-mander, Xander! Xander, Xander,
Xander, bo-bander Banana-fana fo fander Fee-Fi-mo-mander, Xander! Xander, Xander, bo-

## Introduction

*Oh my God you girls have EARNED THIS!* As a much-needed break from this SLOG through science, I've compiled the ULTIMATE FEMALE FANTASY: a chapter made up of lines from all of your favorite lady-books! This has *everything*!

I was a well-educated young lady from Boston with a thirst for bohemian counterculture and no clear plan.[1] He was an animal.[2] Last night [he] appeared wearing suspenders and a darling little Angora crop-top, told me he was gay/a sex addict/a narcotic addict/a commitment phobic and beat me up with a dildo . . .[3] These facts alone make him an unlikely romantic partner for me, given that I am a professional American woman in my mid-thirties, who has just come through a failed marriage and a devastating, interminable divorce, followed immediately by a passionate love affair that ended in sickening heartbreak.[4] Everyone knows that dating in your thirties is not the happy-go-lucky free-for-all it was when you were twenty-two.[5] Newly energized, I gulped the rest of my coffee, brewed another cup for Alex, and took a quick, hot shower.

When I went back into his room, he was just sitting up.[6] He really is very, very good-looking. It's unnerving.[7] For dinner, we had crabs I'd caught off the dock.[8] I am fat with love! Husky with ardor! Morbidly obese with devotion![9] This is why, in fact, I have decided to spend this entire year in celibacy.[10] It wasn't what my doctor ordered, though. My doctor—my gynecologist, to be specific—ordered a baby.[11]

Time, unfortunately, doesn't make it easy to stay on course.[12] The joy of sex[!][13] I'd never imagined myself a mother, never wanted that.[14] In general, research on this topic finds that although young women often report having a strong commitment to both their future career and their future families, they anticipate that combining the two will be difficult and require trade-offs.[15] I was dating this guy for a year and a half. We'd had a few conversations about marriage.[16] [And now] I am sprawled on his chest in the flowery bower in the boathouse, sated from our passionate lovemaking.[17] The bulbs, along with the whole clitoris (glans, shaft, crura) become firm and filled with blood during sexual arousal, as do the walls of the vagina.[18] The home pregnancy test I just took came back positive.[19] There was a tentative little nudge in my womb.[20] And [I] again bare his brother Abel.[21]

1. *Orange Is the New Black*, p. 5

2. *The Bridges of Madison County*, p. 126

3. *Bridget Jones's Diary*, p. 10

4. *Eat, Pray, Love*, p. 1

5. *Bridget Jones's Diary*, p. 10

6. *The Devil Wears Prada*, p. 28

7. *Fifty Shades of Grey*, p. 16

8. *The Girls' Guide to Hunting and Fishing*, p. 9

9. *Gone Girl*, p. 38

10. *Eat, Pray, Love*, p. 1

11. *Julie & Julia*, p. 5

12. *The Notebook*, p. 2

13. *The Joy of Sex*

14. *Twilight Saga: Breaking Dawn*, p. 132

15. *Lean In*, p. 201

16. *He's Just Not That into You*, p. 89

17. *Fifty Shades Freed*, p. 3

18. *Our Bodies, Ourselves*, p. 1

19. *What to Expect When You're Expecting*, p. 19

20. *Twilight Saga: Breaking Dawn*, p. 136

21. Genesis 4:1

CHECK IT OUT!

# Pharmacolgy
# & Medicine

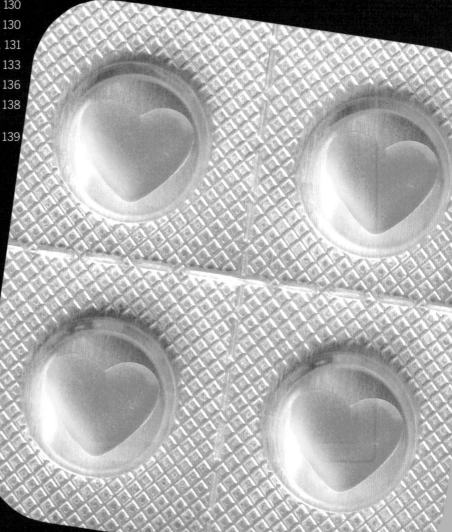

# Introduction

Knock knock! Who's there? Time! What time is it? Time to start another fun chapter!! YAY!

This chapter is coming at the perfect time because, big news: I just got out of a ten-month-long coma, ladies! **FIG. 6.1**

This was probably brought on by my extreme meth-related weight loss, but has the added benefit of contributing to even *more* extreme weight loss! Ladies, there's nothing more tried and true for weight loss with no work than comas!

And as for the pride of dicks (technically more than one dick is called a "pride"): you know the rule—the first girl to fall asleep at the sleepover or to go into a coma gets dicks drawn all over her face. This one's on me! *Literally!* Fun

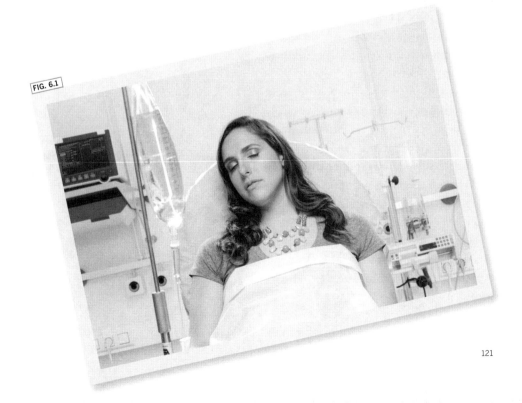

FIG. 6.1

fact: these are all dicks I've previously slept with. Guess which one is Xander's! Guess which one is Waldo's!  FIG. 6.2

FIG. 6.2

In the words of the late Gwyneth Paltrow (R.I.P. 1972–2017 [projected]), "Let's get to the good stuff!" *Medicine* refers to the practice of the diagnosis, treatment, and prevention of disease. *Pharmacology* deals with making and testing drugs to help treat and prevent those diseases. *Pharmacology & medicine* comes from the Wingdings ☐〰〰☐〇〰♍☐●☐♑⊠ 📖 〇♏♎⚹♍⚹◼♏, which, loosely translated, means "square squiggles loopy quotation marks cube sphere loopy quotation marks Virgo symbol other square black circle other square kind of looks like the word 'no' square with caret in it space book space sphere Scorpio symbol Ohm over a line I dunno I guess two parentheses with a line connecting them Virgo symbol that parentheses thing black box Scorpio symbol."

Also, huge news on the X-word front! That's his name, right? "Xander"? I totally forgot because I so don't even care about him anymore. Anyway, I'm pretty sure that Xander (or was it "Sandra"??) came to my hospital room while I was in a coma! Sure, it was just to sign a do-not-resuscitate form, but he really stepped up his game and came. Girls, a tip: always put ex-boyfriends that you kind of want to see again as your next of kin! That way, they HAVE to return your calls! (Or at least the calls of the ICU nurses!) ;)

# Diseases

A *disease* is an abnormal condition that affects the body of an organism. In humans, *disease* is often used more broadly to refer to any condition that causes pain, dysfunction, distress, social problems, or death to the person afflicted. So basically a boyfriend, then, right, gals! :) Can I get a "♌□⊡⚹□♓♏■♎♦•◆♍⚷♋■♎♋□♏♌♍♏•♓♍♋●●⊡⚷◆•♦●♓⚷♏♎♓♍♏•♏♦"?!

There are millions of diseases. If you count ugliness as a disease, there are a million and two (face ugliness and full-body ugliness, a usually more terminal disease). There are only four main types of diseases, though. Just like there are four ladies in *Sex and the City*! You and your three girlfriends can play "Which type of disease are we?" **FIG. 6.3**

**FIG. 6.3**

MAIN TYPES

~~~ OF ~~~

*Diseases*

Deficiency Disease: Charlotte

Hereditary Disease: Miranda

Physiological Disease: Carrie

Pathogenic Disease: Samantha, duh!

*Cancer* is a major illness. It's very virulent—you will 100 percent get cancer if you are a my grandma. There are two types of cancer. One is the astrological sign, and one is a terminal illness, and both are unpredictable! One time one of my friends with cancer just showed up at work with a shaved head and it's just like YEAH you are so unpredictable these days!! Just a crazy rebel who does whatever you want with your hair at any time!! You go, cancer-girl! Straight to the *oncologist*!

hot cancer tip! Ask for a cute oncologist!

Some groups help people and children with cancer, like the *Make-A-Wish Foundation*. The Make-A-Wish Foundation helps thousands of children a year to fulfill a dream. They are a wonderful organization whose motto is "no wishing for less cancer or more wishes."

Make sure you don't have cancer by taking this month's *Science . . . for Her!* featured quiz!

# QUIZ:
# DO YOU HAVE *Cancer*?

*Do you have cancer?*
*Find out with*
*this fun, flirty quiz!!*

1. It's the middle of class and your crush looks over to see you:
   A. Texting him!
   B. Paying attention to the teacher. Come on—it's *class*!
   C. Picking at a large new mole that has recently developed on your forearm!

2. When you're out with your friends, you are:
   A. Gossiping about the cute new boy in fourth-period bio!
   B. Sharing negative cancer test results over some Frappuccinos (pumpkin fraps, come on—it's *fall*, sluts)!
   C. Bleeding from your tumors, you total slut! (Bleeding from tumors = getting to "second base"!!)

3. If you could change one part of your body, it would be:
   A. Your abs and/or abdominals!
   B. Your C-section scar, you postnatal slut!
   C. The lumps you recently found in your left breast, aka Thelma ;) (Finding a lump in your breast = getting to "first base"!!)

4. You and your guy are curling up on a snowy night. What do you do?
   A. Kiss a little, nothing further—you're a good girl, you're no slut ;)!
   B. Go all the way—you're a naughty girl and some would say the "slut" of the night school/Hebrew School :p!
   C. Die!

5. The pop song that most describes you is:
   A. "Pumped Up Kicks" (like ANYONE can resist that great song!! Anyone who says they don't like that song is a total slut, sexually)
   B. "I Wanna Hold Your Hand" (honestly, one of the best songs by one of the best bands EVER, the Beatles! Ever heard of 'em? ;) They don't need to sing about sluts to be so sexy/slutty!)
   C. "My Humps" (referring to the humps/lovely lady lumps in your breasts)

6. If you met a cancer doctor, you'd say:
   A. "Hey, you're a sexy doctor! Wanna listen to Foster the People and touch my Jew-nips?"
   B. "Hey, I'm a good girl but that doesn't mean I'm above being a slut. Wanna touch my Jew-nips to the sounds of Foster the People, this generation's the Beatles?"
   C. "I have cancer, gllrrrrrrrrrrssh" [the sound of blood gurgling out of your femur marrow]

7. When it comes to sports, you:
   A. Are a sports slut!
   B. Are a sports slut but for sex (so just a plain ol' slut)!
   C. Are dead from dying from overdosing on cancer!

8. If your life was an MTV show, it would be called:
   A. *MTV's Slut-Ass Bitch-Ass Slut!*
   B. *Teen Mom's MTV Jew-Nips!*
   C. *Sixteen and Cancerous!*

*Mostly A's*
You go, girl—you don't have cancer! You are also the quintessential flirt, you slut! ;) Also, try wearing "winter" colors like blue, purple, or green. Also, you are a vampire. Also, you are a slut. Also, you're the movie *Harry Potter and the Deathly Hallows: Part 2*. Also, you're an iPad 2. Also, you're a farmhand, you slut.

*Mostly B's*
No cancer, but probably you are a slut! But the good kind! You're a slutty bagel. You're the kind of girl who wears mascara on both lashes which is fairly slutty but foolproof to make boys want to give you the Hoobastank. You're the kind of girl another girl would see and be like "girls are idiots" but really she's an idiot girl too because all girls are idiots and sluts. If you have a crush, you should try texting him a flirty message like "Hello! I am a slut!" Also, you're a Zune, you slut.

*Mostly C's*
You have cancer! You big-jugged slut! You have cancer!

If you're a cool gal (and you all are! MWAH! :)), you've probably heard of *sexually transmitted diseases*, or *STDs*. STDs are a bad way to get guys but a *great* way to prove to the popular girls that you're not a virgin.

One of the most reviled STDs is *Acquired Immunodeficiency Syndrome*, or, for short, *Acquired Immunodeficiency Syndrom*. Acquired Immunodeficiency Syndrom used to be thought of as a certain killer, but now it's nothing more than a slight nuisance, like having a blowout appointment with a new hairdresser who doesn't really know what she's doing and talks too much. All you have to do is take the right mixture of drugs and your Acquired Immunodeficiency Syndrom will be more than under control.

*Diagnosis* is the part of medicine that focuses on finding out what's wrong with you. Sometimes you can look at your girlfriend and know something's up, but you're not sure what. Are they worried about a boy? Is it aseptic meningitis? Are they worried about a *girl*? It's tough to tell. That's why doctors have lots of tools to be able to find what's wrong. Sometimes they put a thermometer up your *butt*! Who knew that taking your temperature could be so *arousing*?

If you're not lucky enough to be able to visit a doctor, you can also use *online diagnostic tools* to figure it out for yourself. *WebMD* is a site where you can input your symptoms and it will output either a diagnosis or a big flashing screen that says "TMI."

**5 FUN WAYS TO GIVE YOUR GUY**

*Acquired Immunodeficiency Syndrom*

1 GIRL-ON-TOP

2 DOGGIE STYLE

3 SIXTY-NINE INTO MISSIONARY (69 → =)

4 SHARING DIRTY NEEDLES TO SHOOT HEROIN WHILE HE'S SPOONING YOU FROM BEHIND

5 REVERSE COWGIRL TAINTED BLOOD TRANSFUSION

It's best just to not get diseases in the first place. We've all heard "an apple a day keeps the doctor away." Well, an apple a day also keeps the *boyfriend* away! Apples have carbs! Blecch! KNOCK KNOCK! Who's there? A fat woman! A fat woman who? A fat woman who can't find a boyfriend because she ate too many apples and now my hands are so fat that my knocking is extra loud which is why the knock knock was in caps!

# SHOTS! SHOTS! SHOTS! SHOTS!

I'm not just talking shots of tequila that you do on spring break in Tijuana and then you pee on a police horse and go to jail! I'm talking *vaccinations*, the shots you get at the doctor. *Vaccines* are biological preparations that improve immunity to a particular disease. They often come in the form of *injections* or *oral administration* (mmm, "oral administration" was my nickname in Mexico jail ;)). Many diseases, like *polio* and *smallpox*, have been cured or almost cured with vaccines. Plus, the disease of horniness can be completely eradicated with a *hot-beef injection*!

Though vaccinations might seem like a no-brainer (like buying that ecru Birkin bag! You treat yourself, girl!), they have met with a lot of controversy over the past couple of decades. Former *Playboy Playmate*, *nude model*, and *actress Jenny McCarthy* claims that vaccines caused her son to develop *autism*, a disorder of neural development with mysterious origins. While there is no scientific evidence that vaccines are dangerous, and while all anti-vaccination findings have been found to be false, Jenny McCarthy has spearheaded an anti-vaccine campaign for many years now. Because of her outspoken beliefs condemning shots, the rate of children being vaccinated against childhood diseases has plummeted and the rates of those diseases have skyrocketed. Measles, mumps, and whooping cough are all on the rise over the past ten years. I used to call Xander's dick "whooping cough," because it was always on the *rise*! ;)

What's the takeaway from all this? Ladies, if this Jenny McCarthy stuff teaches you anything, it's that ALL ANYONE WANTS TO DO IS LISTEN TO A HOT BLONDE TELL THEM EXACTLY WHAT TO DO AND HOW TO THINK. Look at this hot bitch!! **FIG. 6.4**

**FIG. 6.4**

Jenny McCarthy has even less science experience than *me*, an alleged astronaut killer, and she's been able to affect the entire landscape of American public health!! While there is no proof of a link between vaccines and autism, there sure is a link between being a hot lady with big melons and having everyone listen to you!!! Just keep on keeping your body tight, gals, and you can basically make up *any sort of science you want*!! Hot women truly CAN change the world!

# Brain Medicine

Knock knock! Who's there? Me! Me who? *MEOW!!!! CUTE KITTY!!!* `FIG. 6.5`

`FIG. 6.5`

Sometimes your brain doesn't work exactly the way you want it to. It's like a pledge for Delta Gamma that you have taken under your wing but then she's shitty anyway and fucks a guy from Alpha Epsilon Pi even though your brother frat is Sigma Alpha Epsilon!

*Mental illness* is a broad, generic label for a category of illnesses that may include affective or emotional instability, behavioral dysregulation, and/or cognitive dysfunction or impairment. Life can just get to you sometimes, ya know? I myself have dealt with depression before. After Xander and I broke up, I was in a really bad place (Newark). Additionally, I get something called *seasonal affective disorder*, or *SAD*, where I get depressed when McRibs aren't in season. I also haven't always been super wealthy. You might not guess it, but my Tom's shoes are the poor-person pair.

Mental illness is treated with a variety of therapies and medications. Seasonal affective disorder is often treated with the McRib. Unfortunately, sometimes even these treatments fail.

Remember: suicide is never the answer (though *attempting* suicide is a good way to get noticed by that hot EMT!). If you get to the point where you want to commit *suicide*, you should call a suicide hotline. You can also text them a bunch of emoji, if you're less of a talker and more of a texter. Try this: `FIG. 6.6`

`FIG. 6.6`

Love you gals! Stay strong!!

# Sad Libs

*Fill in this Mad Lib (that I call Sad Libs) to create your own fun suicide note for any occasion!*

Goodbye cruel _____. Now that I am done with this life, I am finally _____
     *noun*                                                 *verb ending in -ing*

to the great hereafter: _____.
                             *location*

I am leaving you because I am _____. I cannot endure this _____ _____.
                                  *adjective*                        *adjective*     *noun*

This was no one's fault, except for _____. And honestly
                                          *name of closest friend or family member*

I can't say that the _____ government helped. _____ is a
                    *nationality*                           *name of government official*

_____ _____.
*swear word ending in -ing*    *swear word noun*

I guess there are some things I'll miss after I _____ myself. I'll miss _____
                                     *verb*                     *latest Miley Cyrus single*

and _____ M&M's. But mostly there are things I am happy to be rid of forever.
     *favorite flavor of M&M's*

I'm finally leaving _____, _____, and _____ for good!
            *negative emotion*          *synonym for suffering*         *Guy Fieri*

You have been the best _____ a person could ever have. You've been just like a
                           *profession*

_____ to me. Things have just gotten too bad. Everything has gone from me but the
*family member*

certainty of _____ and _____. You are extremely
              *synonym for despair*         *taxes*

_____ and I _____ you!
*adjective describing Whoopi Goldberg*       *verb you do at a circus*

Before I go, please remember to _____ my _____ and _____ my _____.
                          *verb*     *plural pet*           *verb*     *Arli$$ DVDs*

I love you. Goodbye, _____ and _____! I will haunt
                    *Kurt Cobain's daughter*       *Kurt Cobain's wife*

_____ for _____ years!
  *room of your house*      *number bigger than one thousand*

_____, _____
    *adverb*          *your name*

CHECK IT OUT!

# Food Poisoning

*Food poisoning* is any illness resulting from the consumption of contaminated food, pathogenic bacteria, viruses, or parasites that contaminate food. Bad food can cause such terrifying diseases as *E. coli* and *O. besity*. Food poisoning is one of the worst diseases you can get because it's like a close friend is betraying you. It's like a panda turning on its cub. That being said, if your case of food poisoning isn't too serious, it's a GREAT way to lose weight! And if it is too serious, it's a GREAT way to procrastinate on your housework because you're dead!

A good way to avoid food poisoning is to always check the expiration date before you dig into that Spam cup. Like, if a wine label says "2009"? Don't drink that! It's expired, girl! My high school ex-boyfriend Carter didn't believe in labels, like "boyfriend" or "WARNING: BLEACH." Probably why he's not my boyfriend anymore! Because he's dead!

One big form of food poisoning is *botulism*, aka *botox*, which helps us transition smoothly into . . .

# Paper or Plastic . . . Surgery!

Plastic surgery is amazing. You can make yourself so much better looking and feeling! I like to think of plastic surgery as tailoring your birthday suit. However, girls: only get plastic surgery if you feel comfortable. The meth I did effed up my face, and made it *necessary* to get plastic surgery. I personally had to get a boob job for my deviated septum.

# Big Pharma

In the previous chapter, I started going over the science of drugs. That was all about fun drugs. This section is all about the science of *legal* drugs. They may not be as fun as illegal drugs, but they *are* less fun than illegal drugs!

In the United States (woo, USA! Home of the free and land of the *hotties*! Have you *seen* Channing Tatum lately? WHAT A DILF!), drugs are managed by the *Food and Drug Administration (FDA)*. For a drug to be approved by the FDA, it has to fulfill two requirements: 1) the drug must be effective against what it's treating, and 2) the drug must have been extensively tested on animals and humans. And for all intents and purposes, men count as animals! LOL-er skates! LOL-iver Twist! LOLacaust!

Human testing might sound scary, but it's actually a major part of why drugs work. It's also a good way to make money if you're a little strapped for cash. You get to be a cute little guinea pig!!

Maybe you had a bad day at work or maybe you have a bad case of PMS! Whatever the reason, if you're having a bad day, just check out these pics of animals having makeup being tested on them!

Pharmaceutical companies are sometimes called *Big Pharma* because they have a ton of money and can use it to lobby for what they want in politics. Related: I once dated a guy people called Big Farmer. He was six foot five, around 250 pounds. Loved farming. Why he was called Big Farmer I'll never know.

We assume that drugs are automatically being used to help us, but sometimes the people who make them have nefarious plans. That's why you should *always read the fine print*, or *always make a man who's better at reading read the fine print out loud to you.*

# The Fine Print

Magnoz is a once-daily birth control pill.

Do not use Magnoz if you have kidney, liver, or adrenal disease because this could cause serious heart and health problems, or if you have or have had blood clots, certain cancers, history of heart attack or stroke, or if you are or may be pregnant.

Do not use Magnoz if you are over the age of 35. Also if you are under the age of 35. Only 35-year-olds should use Magnoz. Being any other age will greatly increase your risk of breast cancer, stroke, blood clot, and heart attack. Do not lie about your age. Magnoz will know.

Magnoz doesn't not not protect against HIV or STDs. Is that a confusing warning? Sorry! Let me put it in less confusing terms: if you take Magnoz, you don't not WON'T get AIDS, JK! There you go!

Do not use Magnoz and marry someone of a different race. Just in general. I don't support it. Hope you agree! My favorite part of writing fine print has always been how great of a platform it is for radically conservative racial beliefs! People should really read these warning labels, there's a lot of good stuff about which races should marry each other or not.

Take one tablet by day at the same time every day while in the same place. You're allowed to think different thoughts than you did on the first day but they have to be about the same family of things (e.g., if you thought about one *Friends* episode, "The One with the Late Thanksgiving," on the first day while taking Magnoz, you can think about a different ep, like "The One with All the Rugby." Both great eps.).

During the first cycle of Magnoz, take one light-pink pill daily, beginning on the first Sunday after the summer solstice. Magnoz should not be started at any other time of the year, but especially the third day after Samhain, the pagan festival of Hallowe'en. Take a light-pink pill consecutively for thirty-nine days, then take one and a half light-blue pills. Only split the blue pill in half with the master sword from *The Legend of Zelda*. Any other sword and it will cause immediate gastrointestinal distress. Skip Magnoz every third day. Now, by "day" I of course mean a day on Mercury. That's about 58 Earth days.

Talk to your doctor about Magnoz. And if your doctor is a *Friends* fan, talk to her about "The One with the Worst Best Man Ever"! OMG I am gonna LOL just thinking about it! Love you, Ross!

Some common side effects while taking Magnoz are: headache/migraine, menstrual irregularities, nausea/vomiting, breast pain/tenderness, fatigue, irritability, decreased libido, weight gain, and mood changes.

. . . But the most common side effects while taking Magnoz are: eyes falling out, eyes turning to stone, eyes turning to knives and impaling your brain through your eye sockets, and weight gain.

- Olive oil
- String cheese
- Wheat bread
- plums plucots, etc etc

Oh, sorry! I accidentally started writing my grocery list into this document! Just skip that. No time to erase. Just pretend like I never wrote it and keep reading!

Stop Magnoz if a deep-vein thrombosis occurs, or if the *Friends* episode "The One with the Jellyfish" comes on. Stop everything and watch it. It's the best ep. But yeah, also the thrombosis thing.

On a semirelated note, if you take Magnoz, you are implicitly agreeing with all my views on race and racial mixing. They wanted one of the pills to be light brown instead of light blue and I was like SORRY SISTER that's the color of like a mixed-race woman like Rashida Jones and I do not condone that. I tried to make the pills all white but you need to know which ones you're taking I guess. I was painting them all over with Wite-Out for a while but there were some wrongful-death suits and I am just TOO tired to deal with those pesky pains in the neck!

Tell your health-care provider about all other drugs you take. Some other drugs may make Magnoz less helpful, including:
- Marijuana
- Cocaine
- Opiates
- Barbiturates
- LSD

And, like, if you take those, tell me, too? And like also where you get them? I'd like to get some of those. Do you have a good dealer? I would like some drugs. I hate my life and I need to escape. I can barely see anymore after writing fine-print warnings for fifteen, almost sixteen years. Yeah, you have to write them that small. You don't get to write the warnings big and then shrink them down for the back of the prescription bottle or whatever. I feel like I could bring up the idea of writing normal and shrinking but I am just so exhausted. I can't believe brown people are doing this to me by taking all the real jobs and I have to have this job.

The most common side effects while engaging in an interracial marriage are: sweating, fever, and death.

I hate this fine-print life so much, please kill me.

# The Disease . . . of Love!

Major #doy trending topic here: love is a disease. There's no two ways about it. There *is* three ways about it, though. What I'm saying is, three-ways are a great way to prove you love someone.

If love is a disease, then the cure is *wiener*! Every girl should have a boyfriend. I will repeat this: EVERY GIRL SHOULD HAVE A BOY-FRIEND. As you know from reading my book (oh my G, thank you gals again SO MUCH for reading and buying this book!!! You are the best friends ever!!), I've been going through a little bit of a dry spell in terms of true loves. After my breakup with Xander, I haven't found that Prince Charming I've been looking for. I *have*, however, found a lot of nonroyal civilians to have fun with. Like they say, you have to kiss a lot of frogs before you find a frog who will pay for everything and do meth with you! FIG. 6.7

The silver lining is, after he signed the DNR, I've finally TRULY gotten over Xander!!

I know I said it before, but I was totally lying! Like, I may have been ready to sleep with other people, but NOW I'm ready to LOVE WITH other people. I don't even know who Xander is anymore! Why do I keep typing the word "Xander"??? That's nothing! I PUT THE "X" IN "XANDER" BECAUSE HE IS MY EX! MORE LIKE "DO NOT RESUSCITATE" OUR RE-LATIONSHIP! I apologize for screaming! I'm going to go make myself some throat-soothing tea now! Mmm . . . tea: nature's coffee!

Girls, if you have any boys to set me up with, I'd really appreciate you passing along this cover letter. Maybe you have a cute brother? A single gardener who speaks English? A PEZ dispenser in the shape of a human that I can wiggle so that it looks like a boyfriend? Lady tip: there's not that much of a difference between a rare PEZ dispenser and a boyfriend! They're both hand-some, worth a lot of money, and spit PEZ at you!

FIG. 6.7

# Girlfriend Cover Letter

To Whom It May Concern:

My name is Megan Amram, and this letter is to express my interest in the position of "Girlfriend." The opportunity presented in this listing is very appealing, and I believe that my experience, desperation, and non-FDA-approved scratch-and-sniff tramp stamp will make me a competitive candidate for this position.

The key strengths that I possess for success in this position include, and are pretty much limited to, the following:
- Boobs that have been described by previous boyfriends as "supple," "not weird," and "tits."
- Daddy issues that will work to your advantage.
- Weight is mostly water weight.
- You want to call tampons "lady-plugs"? You go right ahead.
- Never been nominated for an Emmy and lost.
- Killed a bunch of astronauts.
- Nose breather since 1994.
- Basically I would guesstimate 99.96 percent of my body is water.

You will find me to be fun, looking for a relationship, well-spoken, on the market, single, and not not single. My wide breadth of experience/mouth gives you the versatility to place me in a number of contexts. Most romantic dinner of all time? Not only will I pay, I'll pull out your chair for you at the Hollywood Sizzler of your choice! Making out in the back of a movie theater? I'm all for it, other patrons of *Despicable Me 2* be damned! Family BBQ? I can eat more ribs than any girl I know! Mmm, ribs!! Sometimes, I can't help but be like, I want my baby back, baby back, baby back—seriously, I want my baby back! Social Services took it away when I tried to sell it for MORE RIBS!

I believe my educational background has fully prepared me for the position of "Girlfriend." I assure you that my dowry is plentiful and 75 percent livestock-based. I can also assure you that I will not be a leech on your finances. I love to work—after our break-ups, previous ex-boyfriends have been known to describe me as a "working girl." One lovingly referred to me as a "whore." I manage to adeptly pair my career with a love of Hollywood nightlife. The bad news is, sometimes I do go a little party crazy. The good news is, chances are I already have your name tattooed somewhere on my body!

Furthermore, I am confident that I could provide value to you as your trophy girlfriend, by which I mean I could confidently win trophies in the following events:
• Rhythmic choking
• Dueling recorders
• Facial impressions of that mauled lady who got that face implant
• Carbohydrate intolerance
• Murderball

Please see my résumé for additional information and ex-boyfriend references, though I would avoid calling the first four (Eric Stone, Sam Linden Jr., Theo Wilson, Sam Linden Sr.), since they've been known to lie about my Jewess's sideburns, my penchant for crying while eating, and the communicability of my totally like obviously non-communicable paunch-rash. Frankly, I find "crying while eating" to be such a misogynistic phrase. I prefer "eating while crying."

I hope that you'll find my experience and interests intriguing enough to warrant a face-to-face meeting, or at least a face-to-ass you-checking-out-my-ass. I can be reached anytime in Hollywood via my Hollywood cell phone, (555) 555-5555. You may recognize the area code as that of Hollywood, where life is glamorous and mostly AIDS-free! Thank you for your time and consideration. I look forward to speaking and/or second-basing with you about this relationship opportunity.

Sincerely,
Megan Amram
(555) 555-5555
Hollywood

# "Legitimate Rape"

Babes, I hate to do this, but let's get serious. Being a woman in a modern world that is physically and sexually violent is so scary. But it is also my responsibility to take on the tough issues. I've tackled religion, I've tackled atomic warfare, and now I have to tackle rape. *Rape* is the crime of forcing another person to submit to sex acts, especially sexual intercourse.

Todd Akin, a longtime anti-abortion activist, served as a Republican member of the House of Representatives for Missouri's Second Congressional District from 2001 until 2013. On August 19, 2012, Akin asserted the following on television: "First of all, from what I understand from doctors, [pregnancy from rape] is really rare. If it's a legitimate rape, the female body has ways to try to shut that whole thing down." FIG. 6.8

I wish this were a joke. But legitimate rape is a real phenomenon (and also the sex move of this chapter, PLEASE don't try it, though, it's only symbolic). Knowledge is power, ladies. You have to understand this so you don't unknowingly spread lies!

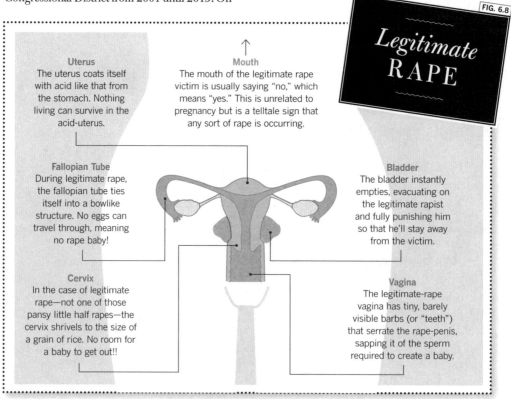

FIG. 6.8

*Legitimate* RAPE

**Uterus**
The uterus coats itself with acid like that from the stomach. Nothing living can survive in the acid-uterus.

**Mouth**
The mouth of the legitimate rape victim is usually saying "no," which means "yes." This is unrelated to pregnancy but is a telltale sign that any sort of rape is occurring.

**Fallopian Tube**
During legitimate rape, the fallopian tube ties itself into a bowlike structure. No eggs can travel through, meaning no rape baby!

**Bladder**
The bladder instantly empties, evacuating on the legitimate rapist and fully punishing him so that he'll stay away from the victim.

**Cervix**
In the case of legitimate rape—not one of those pansy little half rapes—the cervix shrivels to the size of a grain of rice. No room for a baby to get out!!

**Vagina**
The legitimate-rape vagina has tiny, barely visible barbs (or "teeth") that serrate the rape-penis, sapping it of the sperm required to create a baby.

Check out the top mis"conceptions" about rape below! The "conceptions" is in quotation marks because you can't conceive during legitimate rape.

# Mis"*conceptions*" about RAPE

MIS"CONCEPTION"

**THE TRUTH**

You can conceive a child during a rape.  If it's a legitimate rape, the female body has ways to try to shut that whole thing down.

Women who are raped were "asking for it" based on their provocative dress.  It doesn't matter what you're wearing. Being beautiful is asking for it! If you truly didn't want to be raped, you would gain forty pounds and/or come out as a lesbian. Being a beautiful/average-looking woman is a known temptation, and that's on you.

Everyone knows when a woman says no, she often means yes. Women secretly want to be raped.  This is a horrible fallacy. In truth, it's no *secret* that women want to be raped! It's a pretty safe assumption! Most will tell you at length how much they "don't" want to be raped, but it's just sarcastic!

# Nobel Prize in Medicine

Some little girls dream of winning an Oscar. Other, uglier little girls dream of winning a Nobel Prize. The *Nobel Prize* is the highest level of achievement that someone can win in the field of medicine. And how cute would a field of medicine be?! Like, the grass would be made out of little *stethoscopes* and the flowers would be *bags of blood*! OMG so cute!

In 1962, *Watson, Crick, and Wilkins* were awarded the Nobel Prize in medicine for discovering the structure of DNA. Though she was an integral part of the discovery of nucleic acids, scientist *Rosalind Franklin* was never recognized, almost undoubtedly due to the fact that she was a *woman*.

However, this alleged "sexism" is not what we care about. What we care about is the behind the scenes! Who was she wearing! What were the *snacks*! FIG. 6.9 ←

CHECK IT OUT!

FIG. 6.9 ← AGAIN! NICE!

### BEHIND THE SCENES
## *Rosalind Franklin's Lab*

Rosalind can't discover DNA and have it stolen from her without all her *fave snacks*! Roz nibs 'n' noshes on peanut M&M's (on cheat days!) and celery (when she's being a good little girl!).

Snooze! DNA research can be so *boring*! That's why the Wizard of Roz brings *Sudoku* to play with while she's working. Only Mondays, though! Fridays are *way too hard* for this little girl!

I ♥ my job!

Rosalind loves keeping her work in its place (except when it's stolen!). To organize all of her notes and papers, Rozzo uses the *Post-it Desktop Organizer* in dazzling pink! Every girl needs a little pink in the workplace!

Sometimes Rozzie Rozbourne forgets to think and daydreams about making spinach soufflés in little ramekins. That's why she brings her *drool catcher* to make sure she's not drooling all over the place when she's being a dummy!

# Pharmacology & Medicine Recap

I hope you enjoyed this chapter as much as I enjoyed writing the parts of this chapter that I wrote while I was on Ecstasy! FIG. 6.10 If I can save one

FIG. 6.10

girl's life with information about medicine and health, then I've done my job. I'm the Johnnie Cochran of science textbook writers!

AND IN OTHER BIG NEWS, Rihanna lovers: my girlfriend cover letter has been disseminated all over America! I'm pretty sure by the time we come back for the next chapter, I will have found my true love. Aka I will hold off on writing the chapter until I find my true love. Wish me luck!

# Recap Questions

**QUESTION 1:** How many people have you sent my girlfriend cover letter to?

**QUESTION 2:** Why haven't you sent it to more people?

**MAD LIBS QUESTION 3:** What is the most _____ _____?
                                              adjective          noun

**QUESTION 4:** Does "no" mean "yes"? (This is a trick question: either way you answer, it means "yes"!)

**QUESTION 5:** Should you talk to your doctor to learn more about Magnoz (SPONSORED QUESTION)?

YOU'RE KIDDING ME!

# 7

# Space & Technology

# Space & Technology: An Introduction

NO NO I CANNOT GET STARTED ON SPACE & TECHNOLOGY BECAUSE I HAVE HUGE NEWS, GIRLS!!! I have to start with that!

# My Big News: An Introduction

OMG I can't believe I'm telling you this right now, for real! Wait, though—I should be mature about this. I am an *author.* I should do my job first and then tell you the news.

# Space & Technology: An Introduction

So um well, "space & technology" comes from the Greek . . . wait, who am I kidding! There's *no* way I'm going to be able to concentrate on this boring stupid old science stuff while I'm sitting on this huge news! Oy vey! I'm so sorry to have jerked you all around but let's try this again!

# My Big News: An Introduction

No wait I am being so bad. I should seriously just do the science stuff. I am being paid for it, and that's money that is funding my hankering for pretzel M&M's that I get around my period (how can I NOT be bad around my period?! :p). Not that I'm just doing this writing stuff for the money, I really truly do love you girls. :) But yeah, let me do the science stuff.

# Space & Technology: An Introduction

F%<# this science WE GOTTA DO THE NEWS FOR REALSIES.

# My Big News: Introduction:
# For Realsies

I AM DATING A GUY!!!!!!!!!!!!!!!!!!!!!!!!!!!!!!!!!!!!!!!!! Like *seriously* dating a guy!

I am sobbing right now, FYI. I can barely talk about my guy without crying. I am just that lucky. I am ajcryinsg so ahrd that teh keabyoard iss geitng so slipeiery taht its hard to tpye.

Okay, I've wiped off the keyboard and have taken a step back, which makes it a little harder to type but it's worth it because I'm not crying on the keyboard anymore. It turns out that I really just had to let Xander go FOR GOOD and it would immediately open me up to a new boyfriend!

We met in the hospital. We were both in clinically induced comas, but apparently even when we were both unconscious, our hands moved a little toward each other. The orderlies at the hospital noticed this, and soon they were dressing us up to go on "dates," "prom," etc.

They even made a calendar of us that you can buy online through Etsy! All the proceeds go to worthy causes like "CUTE HATS FOR COMA DATES," a charity that provides hats for people falling in love during their comas like us.

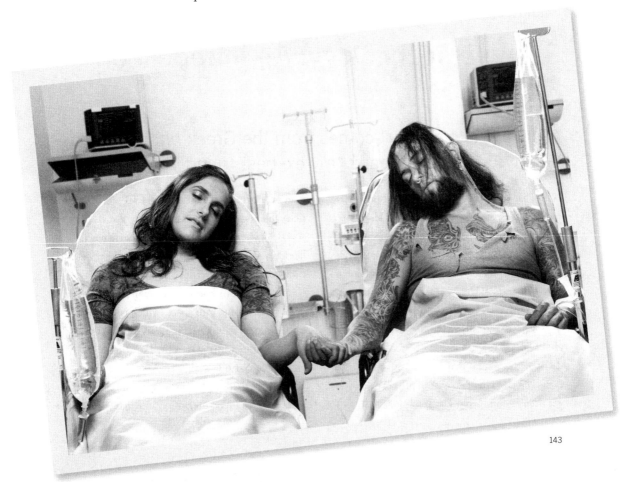

Anyway, after we woke up, we REKIN-DLED the hospital romance! My boyfriend didn't do so well in the coma, TBH—he can't remember most things and I have to bathe and feed him, but is that really that different from most guys? ;) He doesn't really have a name, because trying to call him any one name will weirdly send him into a violent rage, but is that really different from most guys? ;) Can I get a "NO IT'S NOT YOU AND YOUR BOYFRIEND HAVE A PERFECT HEALTHY RELATIONSHIP, MEGAN"?! Thank you!! Great—you gals are so sweet thta Imm cairying agiang!!!!

# Space & Technology: An Introduction For Realsies

## "Space & technology" comes from the Greek *bitch*, which, loosely translated, means "my ex-best friend Tiffany."

When we say *space*, we're talking about *outer space*. And when I say *we*, I'm talking about big, beautiful babes in America. Outer space is the void that exists outside of the earth's boundaries and between other *heavenly bodies*, like the *sun* or *Gisele Bündchen*. Um, *yes* Gisele has a heavenly body!! Love you, Gis!

Space is mostly a *vacuum*. We sure know about those, ladies! But don't be fooled, it's not the type of vacuum that makes your life worth living and gives you the sense of purpose that you get from cleaning your family's house that your man owns. It means that there are very few particles floating around and it's mostly just empty space. Maybe a few particles of hydrogen, maybe like a cupcake that a cute, single astronaut girl was eating on a space date with the moon! OMG how cute! Does anyone know if the moon is single? ;)

There's no air in space, which is okay because humans can live without air for like, three years or something. Or wait, I'm thinking of changing your oil. Yeah wait, you CAN'T live without air, that's it. No air in space means that you can't live there, no matter how many single moons there are to date. Honestly, it's crazy how fragile humans are. It's like, how much air can you possibly breathe?! Save some for the land-fishes! FIG. 7.1

Astronauts are the cool dudes who are trained by a human spaceflight program to command, pilot, or serve as a crew member of a spacecraft. As we learned before (unless you, like my unnamed boyfriend, have massive memory loss and have forgotten the last chapters of this book!), women *can't drive.* Therefore, it makes sense that most astronauts have been men. It's just like that old childhood rhyme: "Girls go to college to get more knowledge. Boys go to Jupiter because they're in control of NASA, science, and their emotions."

FIG. 7.1

# HUMANS
## FRAGILE, YET RESILIENT
### *an Anecdote*

66 *Sometimes the human body is very resilient. I once left a baby in a car with rolled-up windows for like five hours. I'm not an idiot, though, I left the car running so that the baby could drive to a water park if it got hot. I wonder what that baby's up to. FYI, that "once" was five hours ago. The baby's still in the car. And don't worry. It won't get too hot. The car's in my garage!* 99

# The Big Bang

The *big bang* was how the whole entire universe started! Yes, everything: the plants, the trees, my best friend Maddie, my best friend Lizzy, my best friend Sophie, my best friend Anna, Khloe Kardashian (not my best friend . . . *yet*), chard, a baseball card, Mandarin Chinese, hand lotion, face lotion, esophagus lotion (that's what I call yogurt!). According to the big bang, the universe began approximately 13.798 ± 0.037 billion years ago when a tiny little dense chunk of matter exploded and rapidly expanded into all the matter we now know as everything in the universe. It's like when you eat a pretzel M&M on your period and suddenly you immediately balloon into a fat cow! That's what the big bang was like!

But you know what big "bang" makes me think of. I don't even have to tell you! You totally get it because you're one of those cool girls who loves sex but ALSO knows her place! **FIG. 7.2**

There are many other theories about how the universe was created. Some people believe that God created the universe and that there was no big bang. Like most scientists who believe in the big bang, I believe in the big bang. But I can see the other side, too. If God *didn't* create everything with intelligent design, then how is your finger the *exact* right size to put in your man's butthole? It's very confusing!

---

**FIG. 7.2**

## *Tips for Hosting Your Own* BIG BANG

Is your orgy getting out of control? Follow these fun and flirty tips to keep your "big bang" in check! (THIS CHAPTER'S SEX MOVE!)

**Always have enough snacks.**
A hungry orgist is a crabby orgist.

**Set the mood.**
Light lavender-scented candles!
Note: no fire too close to buttholes.

**Serve roast chicken.**
Gotta keep up everyone's energy with lean protein!

**Background music!**
Is it a happy orgy? Try "Walking on Sunshine"!
Is it a brutal demon orgy? Play the music from the ass-to-ass scene from *Requiem for a Dream*!

# Planets

A *planet* is an astrological object orbiting a star. Planets are divided into *solar systems*, which include some random number of planets orbiting a sun. Our planet is called *Earth* and our sun is called *sun*. My favorite planet is Earth, because that's where malls are! FIG. 7.3

FIG. 7.3

Earth is the third-closest planet to the sun, which is why it's so sunny in the summer! A year is 365 days because that's the amount of time it takes for Earth to orbit around the sun. Conversely, a period is about five days because that's the time when the moon is close enough to Earth to *pull your uterine lining* out through your vagina like a *vagina tide*. Gravity is *very* important!

Earth is about 4.54 billion years old. That's so old! That's like a million years old! Earth would be able to orbit the sun until the sun begins dying. In around five billion years from now (August 11 at 3:33 P.M. PDT, I'm sponge-bathing my boyfriend and wearing a child's bathing suit), the sun will become a *red giant* (more in the star section) and expand to engulf Earth in flames. Now *that's* what I call a *burning sensation*!!! That's not what my doctor calls a burning sensation, apparently—he thinks coma boy gave me "something." It's like, tell me what you *really* think, doc! FIG. 7.4

FIG. 7.4

There are seven other planets besides Earth. They are: *Mercury*, *Venus*, *Mars*, *Jupiter*, *Saturn*, *Uranus*, and *Neptune*. There used to be a planet called *Pluto*, but it was kicked out of the solar system in 2006. This is basically the first case of *interstellar bullying*. What did Pluto do except chill and be a sweet li'l guy?! Just because he's a little smaller and nerdier than the other

147

planets, Pluto gets kicked out and can't do any of the fun stuff with the other planets. I'm talkin' planet prom, planet farmers' market, etc. etc.

You can reach the other planets using *spacecrafts* like *rockets*. Maybe you've heard of Pocket Rockets™? Aka my fave type of vibrator, ladies! Rockets are like Pocket Rockets™, except that they go into the sky and out into space! Think of space rockets as the Pocket Rockets™, and outer space as the vagina!

You may have heard about Mars and Venus—which are both *planets*—from the book *Men Are from Mars, Women Are from Venus.* That book is a liar! It should be titled *Men Are from Earth, Women Are Also from Earth.*

Now, I have a bone to pick about this next topic, *the moon.* So many people truly hate the moon, and I just don't get it! I mean, like 90 percent of the people (and women!) I meet, the first thing they say to me is—and pardon my French— "Putain, comment pouvez-vous aimer ce morceau de merde la lune, you fuck????" FIG. 7.5

I guess my beautiful French friends are just negative these days. Even beautiful, majestic orbs like planets or moons or breasts can have their detractors.

FIG. 7.5

# The Moon

**ALL COMMENTS** (13)

Top comments ⌄

**Fatherprick** *1 day ago*
Stupid fuckin color (grey wtf is that even a color?!?!)

**SitDownOrEatKnife** *1 day ago*
Stupid fuckin shape (round, wtf)

**MrMEGAmotherFUCKER** *1 day ago*
Too big didn't look

**TakeYourBath** *1 day ago*
Dis moons gay

**FangedWang** *1 day ago*
Whoever commented dis Ur a Fag

**AreolaPuffington** *1 day ago*
Thumbs up if u were sent here by Tosh

**Mom69** *1 day ago*
Stop making fun of the moon! This is the moons mom and I just want to tell you that the moon is so lovely and is trying so hard and for you to come here like this and make fun of the moon is truly sick. If you'd like to talk more about the moon, you can reach me at (624) 134-9420, thank you, love you sweetie

**suddenlykitties** *1 day ago*
I thought it was ok no homo

**Pat Magroin** *1 day ago*
Fag yes faggo

**MolesterStallone** *1 day ago*
Obama's moon isn't MY president's moon where's its birth certificate

**BAMyouhaveaids** *1 day ago*
It's the 21st century and we still only have one goddam moon????

**injailumyho** *1 day ago*
Lol moon ur moms a fag

**Mom69** *1 day ago*
Ur a fag! Love mom

# Stars

Stars are a huge part of most women's day-to-day life! We love to read about stars' lives! We make decorations on cupcakes in the shape of stars! Our ankles have tattoos of stars! Our lower backs have tattoos of stars! Our faces have tattoos of teardrops! We killed another woman in prison! My friend Maggie has a dolphin tattoo on her lower back. I thought it was just for fun, but I guess she killed a dolphin and the tattoo law for that works similar to the human/teardrop thing? What I'm saying is, any woman you see with a dolphin tattoo has killed a dolphin.

On a semirelated note, a *star* is a massive, luminous sphere of plasma held together by its own gravity. They don't look like that cute little star shape that we see in the cartoons—they're actually more like spheres. Stars are all "apples," huge waists, no ass. Here's FIG. 7.6 what a star would look like if it could lose weight:

FIG. 7.6

---

## *How to Get the Stars' Hot Looks!*

 **1**

Powder your face with orange and/or red costume foundation!

 **2**

Once you have a smooth undercoat, streak it with red eyeliner to get the "veins" of stars.

 **3**

If the coloring isn't working, punch yourself in the face a bunch of times! The slight bruising should give you the color you need to approximate a supernova.

 **4**

If punching yourself in the face doesn't work, find a boyfriend or husband and irritate him enough that he punches you in the face! While many men claim to be "anti" punching women, you can definitely send them over the edge with a well-placed gab.

 **5**

Are you in too deep? You gave your husband/boyfriend/father/teacher a taste of hitting you, and now he can't stop? Well, check out what you're wearing! Are you asking for it? If so, change IMMEDIATELY!

 **6**

Has he progressed to burns? Well, CONGRATS—that looks way more like a star than just a punch! Stars are balls of burning gas, so this is going to be great for your realistic star look. Think of stars as very fat, very round women who have angered their husbands/boyfriends such that they have burned them all over for millions of years!

169

CHECK IT OUT!

# Astrobiology

Earth is the only planet known to harbor life, but there is a lot of research going into finding life on other planets. We call life outside of Earth *extraterrestrial life*, or *alien life*. Alien life is different from just aliens from Mexico or whatever, though maybe they'll be as good at being maids as the aliens from Mexico! I honestly don't know what I would do without my maid Pilar, who is one of my *mejor amigas* and DEFINITELY a naturalized citizen WINK WINK. Like I always say, cleanliness is next to godliness which is diagonally across from Mexicanliness (LUV U, PILAR!). `FIG. 7.7`

`FIG. 7.7`

The best part of finding intelligent life in space is that we'll *finally* have competition for Miss Universe! I'm sick of all these Earth bitches winning year after year. Sorry I called them bitches! I am honestly so happy every day with my coma-man ("coman"???) that I don't know why I even use the b-word anymore. The only b-word I'm going to use from now on is "bathing sponge"! That's when my coman and I are the closest, when I'm softly sponge-bathing his bedsore scars. ;) It's like, ever heard of manscaping, guy?! Use a little scar cream, babe! Hahahahaha we're in LOVE!

When we *do* find intelligent life on other planets, you're going to want to be first in line to date them. And hey, if it's *intelligent* life, that's still better than most male life on Earth! Haw haw! My guy has a post-coma IQ of 23, but a DICK-Q of 10 . . . inches! Read this next segment to hear the pros and cons about dating the microbes that scientists recently found on Mars.

---

## *Carbon* DATING

**Scientists recently found microbes under the uninhabitable soil of Mars. And I bet they're *cute*!**

**PRO:**
Microbes are up for anything! They'll never complain about accompanying you to a wedding, because you can put them in your pocket and just take them there.

**CON:**
They're very small, which means they've probably developed aggressive and/or annoying personalities to overcompensate.

**PRO:**
Found living in dirt, which means they won't have high standards for how clean Pilar keeps your house.

**CON:**
No dicks.

**PRO:**
No hair.

**CON:**
They fucked my roommate last year.

# E-mail

Let's transition smoothly into the technology section of this chapter!

Transition accomplished! I am so happy to be living in the time that I am. A hundred years ago, they didn't have any cool technology, like computers or calculators or 2014 editions of the Zagat guide! I know they say the economy is bad, but sales of 2014 Zagats have never been better.

Before we get too deep into technology, I'd like to issue a small warning. Technology does not equal safety. As a young modern woman (yes, babes! Can I get a "ROAAAAAAAAAAAAAA AAAAAAAAAAAAAAAAAAAAAAAAAAAAA AAAAAAAAAAAAAAAAAAAAAAAAAAAAA AAAAAAAAAAAAAAAAAAAAAAAAAAAAA AAAAAAAR"?!), 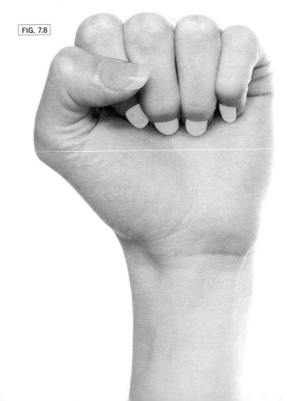 **FIG. 7.8** we have to be careful everywhere at all times. Not even your e-mail is safe from hackers and spammers. *Spam* tries to hack into your accounts so it can get your credit card numbers (Can I get a "4766 6359 1249 5985 Exp. 03/17 CVV 476"?! That's my favorite credit card number, it's SO CUTE!) and buy shoes for itself.

*Spamming* is the use of electronic messaging systems to send unsolicited bulk messages, especially advertising, indiscriminately. Spam has been around before the Internet was even invented.

**FIG. 7.8**

Save As Draft   Photo Browser   Show Stationery

Subject: URGENT REQUEST FROM MOOR OF VENICE

Signature: Signature #1

### URGENT REQUEST FROM MOOR OF VENICE

O Hello ! I am a Moorish prince ! It is with heart full of hope &tragedye that I explain this tragedye.

my wife Desideminna was killed with a stab &and I tragically cannotget in her will which left me many of her possessions: moneyes, whitescarves, whiches bramble. Please help me live wi/out the brambles by donating your pence !

I hereby agree to compensate your sincere effort with 20% of the funds, pay'd back in puffed crevats, greensleeves sheet musics&slaves. no risk no danger.contact my barrister @: t8wrguhisd-upon-Avon.

Best wishs,
Othello T. Smiths


### CONGRADULATIONS FROM THE CAPULETs!!!!!!!!!!!!!!!

Your eMAIL (which of corse means "ENGLISH MAIL" ) has been selected by the board of the Catapult household for entrance to our Romero&Juliette ~DREAM BALLET~.

only 14 lucky WHITE ACTORS have been invited.to attend, please provide :

1. FULL NAMYE: ..................
2. RESIDENTIAL ADDRES: ................................
3. DATE OF BIRTHE: ............................
4. PIN NUMBER (NUMBER OF PINS YOU OWN): ..............................
5. DATE OF FIRST CHILD'S DEATHE: ..................
6. NUMBER OF TIMES YOUVE HAD THE BLACK PLAGUE: ..................
7. FAVORIT TYME YOU HAD THE BLACK PLAGUE: ..................
8. OCCUPATION OF FAVORIT SLAVYE: ..................
9. QUEEN: ..................
10: FACTS NUMBER (NUMBER OF FACTS YOU KNOW ABOUT PINS): ..................

Thank you for your cooperation,
Lady Cpulat G.F. Aziz Bello


### Much ado about LADYES !

Ladyes ladyes LADYES !

See why thousands of LORDS have checked out LADYES

all ladyes !
barely-clad ladyes !
heaving bosoms withheld under but FOUR LAYERS of brocade !
ladyes with gout !!

- - Send but thy address & 140000 pence & 1 slavyes & your PayPal (friend who payes 4 you when ur in town drinking ayle) to Richard the 5494jkdsfh8ith at Hereford Avenue, Wales

**WANT TO GET THAT DAMN SPOT OUT FAST??????**

~*~LOSE ONE DAMN SPOT IN AS LITTLE AS TWO DAYS~*~

don't wastye you're time with scams that promisye SPOT-LOSS for LITTLE OR NO EFFORT. our patented revolutionary spot-busting techniques guaranty results !!!!!!!!!!!!!!!!!!!

~*~MAKE YOUR THANE BE LIKE DAMN MY LADYE BE LIKE THE SUN WHEN SHE POWDRES HER TEATS~*~*~*~

you will LOSE YOUR DAMN SPOT
you will have IMPORVED CONFIDENCE, SANITY, & TEATS
this is the BEST DAMN SPOT SHOW PERIOD

Pease send Send 9999999999fdgsd9f999999 SLAVYES 2 LONTON THIS IS NOT A JOKE THIS IS FOR MY DAUGHGTER

Thank you!!!
Ladye Macbeth 8. Chang


URGENT REQUEST**:
SHALL iCOMPARe THEE 2 A SUMMER'S DAY????????????????????
~*~*~*~*~*~*~*~*~*~*~*~*~
Send 2545375246y6 WEST INDIAN SLAVYES 2 CARDIFF O YES PLEAS


**CAN I TRUST YOU?????**

Dear friend,

I know this letter will definitely come as a surprise to you but I hope

you will keep reading

No, Thank YOU,
Dr. HenryIVpart2


**VENCEDOR - notificaзro final::**

SEU EMAIL (ENGLISH MAIL) ID ganhou (₺ 450.000,00 ha'pennies) nas competiзxes internacionais de espanhol "El Gordo" loteria Email prкmio,

nъmeros da sorte e 9/11/13/24/43 Ref: ES/9420X2/68. NOTA: Este й um programa de loteria internacional, o aviso tenha sido traduзro de Inglкs para Portuguкs, porque vocк й um vencedor. O contato Esclarecimento e procedimento: GARANHГO AGКNCIA RECЛAMAЗГO MEGA

LOVE,
hamlet

# E-male!

Men are on the Internet, too! Though it sometimes feels like women are the only ones who are single these days, there are plenty of dating sites. It can be overwhelming. When should I use Ok-Cupid, and when should I sit on my butt eating a pint of some fun Ben & Jerry's ice cream (the Passion Fruit of the Chrust, which is passion fruit and the crusts of French toast)? Use this fun, flirty quiz to find out!

*Which*
## DATING SITE
*is*
## RIGHT FOR YOU?

Take this quick fun and flirty quiz to see which dating site is right for you!

**Which dating site is right for you?**

**A.** OkCupid

**B.** Match.com

**C.** Tinder

MOSTLY A's: OkCupid
MOSTLY B's: Match.com
MOSTLY C's: Tinder

Before you meet someone off the Internet—BE CAREFUL! More than 70 percent of women currently will be murdered by someone off of Craigslist. Are you in public? Look to your left. Look to your right. Both those people will be murdered off Craigslist. By you! Take control of your destiny, bebs—not just men can be serial killers! Don't let anyone tell you you can't kill the two people next to you! As someone with a healthy image of mortality, I think I'll be great at killing someone when I feel ready. Your first time is very special! Don't use it up on just *anyone*!

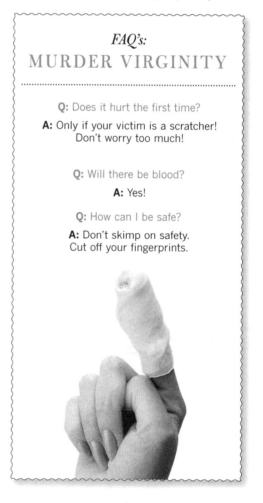

*FAQ's:*
## MURDER VIRGINITY

**Q:** Does it hurt the first time?
**A:** Only if your victim is a scratcher! Don't worry too much!

**Q:** Will there be blood?
**A:** Yes!

**Q:** How can I be safe?
**A:** Don't skimp on safety. Cut off your fingerprints.

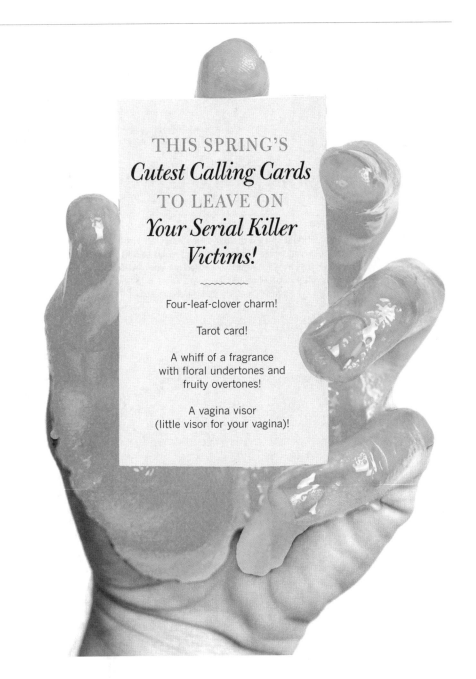

THIS SPRING'S
*Cutest Calling Cards*
TO LEAVE ON
*Your Serial Killer
Victims!*

Four-leaf-clover charm!

Tarot card!

A whiff of a fragrance
with floral undertones and
fruity overtones!

A vagina visor
(little visor for your vagina)!

Remember that going out on Internet dates can be VERY dangerous. You have no idea if they're dangerous criminals or ugly! Those two things are equally bad! Take all precautions to protect yourself, goils. FIG. 7.9

FIG. 7.9

# Sexual *Assault* ... and Pepper! ... Spray!

**Every woman should carry a pepper-spray canister with them at all times, because you never know when you'll need to add spice to a recipe that's just not cutting it! Or protecting yourself from a guy who's about to assault you or whatever. Here are some of *Science ... for Her!*'s favorite recipes using pepper spray.**

### SPICY LIME COCKTAIL
Mix equal parts vodka, lime, simple syrup.
Spritz with pepper spray.

### MANGO-PEPPER SPRAY PALETAS
Freeze mango juice in popsicle molds.
Dust with pepper spray and lime zest.

### FRAGRANCE
Not necessarily a recipe, but just as fun! :)
Spray a little pepper spray under your arms and neck and face. Your boyfriend will HATE it! :)

Now, I know more than anyone (except for maybe Shirlene, my best friend who has never dated anyone and never will because she isn't confident enough, I tell her that all the time) that it's hard to find the right guy. Not all of us are lucky enough to have slipped into a life-threatening coma and gotten set up with our soul mates while on a breathing machine. If you don't find your guy right away, you might want to freeze your eggs. FIG. 7.10

FIG. 7.10

# FUN WAYS *to* FREEZE YOUR EGGS!

**1.**
Throw those eggs in a chilled Martini-ini™ (martini with a bikini coozie around the glass)! Make sure your martini glass has its beach bod ready!

**2.**
Freeze them in a huge block of ice and sculpt it into an ice luge! Send some Absolut™ vodka down your EggLuge™, or, for a little fun, try oySTARS™ (oysters + starfish meat rolls) instead!

**3.**
Take them in a little FantaPack™ (fanny pack to carry Fanta) on a cruise to Canada™ (Canada was recently purchased by China which now owns the trademark)! Brrrrr! Chilliest place on Earth™ (Earth presented by China)!

**4.**
Live in Finland™!

I can't talk about men and the Internet without bringing up *pornography*. Pornography is defined as printed or visual material containing the explicit description or display of sexual organs or activity. It's apparently a huge part of technology. I've never actually seen porn on the Internet, but I believe it exists. It's like God—you don't have proof that He exists, but you have to believe it and *self-flagellate* until you have righted your wrongs. Look at how good I've gotten! I can flagellate myself a tramp stamp! FIG. 7.11

FIG. 7.11

Here's what I imagine porn looks like, based on the hearsay and rumors that float around.

# Economic Technology

*Economics* is the science of the production, distribution, and exchange of goods and services. Can I get a "$$$"? (Pronounced "*Sssssstchkzh.*") I love money! That's what I use at malls! Money comes in different forms: *cash, coins, credit cards.* I've always been fascinated by credit cards: how do they fit all that money inside such a tiny credit card? Answer: *nano-tubes*!

*Political economics* is different from me balancing my checkbook after going on a shopping spree. It's how an entire country balances its checkbook after going on a shopping spree (which is one of my top two types of sprees, right before "killing"!). The United States of America has to handle the economics of 314 million people (aka 157 million eligible bachelors, that's what I'm talking about, babes! Plenty of fish in the sea!). It's not easy to finagle, especially since America has been hemorrhaging money in the twenty-first century. That's why the United States has turned to the crowd-sourcing fund-raising site Kickstarter to fill in the cracks!

CHECK IT OUT!

# National Debt

by the United States Government

**82**
backers

**$2,619**
pledged of $200 billion goal

**4**
days to go

## ABOUT THIS PROJECT:

Hi you guys! Joe Biden and the rest of the gang here! :) We're looking for some awesome people to help us Kickstart our dream project of having a functioning federal government! That's where you come in: all we're asking for is a little help. And twenty trillion dollars.

As you may know, we (the United States government) are a little strapped for cash. Salvage a first-world government's economy? In *this* economy?! As the kids say, "LOL!" (Laugh On Line!) We may be the ones responsible for "this economy" in the first place but still. Uncle Sam may have gotten us into this mess, but WE WANT YOU . . . to GET US OUT!

There is little if any funding available for small-to-midsize debt-based projects such as this. Through Kickstarter, with your support, the country that you live in can remain a free sovereign nation instead of having to sell Ohio to China, 'cause then Ohio would probably start speaking Chinese, and that's *FUCKED UP*.

## A LITTLE BACKGROUND:

For those of you who don't know, the USA is the best! Originally from England, the United States government has been a major world power since it was founded in 1776. The US has brought you such great things as sugar, mistresses, and obesity. Proud home to milk and Ashton Kutcher. Really into righteous wars!

For those of you into civil rights: no slavery! For those of you into slavery: we used to have slavery!

Imagine that famous picture on the cover of the *Sports Illustrated* Swimsuit Issue of Terri Schiavo

**PLEDGE $10 OR MORE**
For your fairly useless donation to help the United States of America not founder under the multi-trillion-dollar debt that we have amassed over decades of misspending and unnecessary wars that some may argue constitute war crimes, you get a tote.

**PLEDGE $20 OR MORE**
If you donate $20, we will list you as an associate producer of the government by carving your name into the Vietnam War Memorial. You can tell all your friends you're a ghost because you died in the Vietnam War! You know—the righteous one!

**PLEDGE $25 OR MORE**
I didn't want to say this to the $20 people, but those guys are assholes. What kind of an asshole only donates $20 to a multi-trillion-dollar debt that is growing by $4 billion every day and has no sign of slowing? Twenty-five dollars, now that's the MONEY-money! If you donate $25, you get a tote (large).

**PLEDGE $50 OR MORE**
You've finally taken the responsibility of the country into your own hands! I will make you a tie-dye T-shirt and cook you and five of your friends a hot dog BBQ at my gf's place.

**PLEDGE $100 OR MORE**
If you pledge $100, which so far only my mom and Barack's mother-in-law have done

holding an American flag. That's *our* flag! On second thought, I think that was a painting I did in 2004 after doing grass for one of the first times. Still, she looks great in a tankini. As the kids say, "LOL!" (LOL A Lot!)

**THE PROJECT:**
We'd like the United States to be fiscally autonomous. It's been in the works for many years now, and we think it could be great. Thanks to Kickstarter, we have a chance to reach individuals who will personally bail us out of this mess. How great for us!

I know it's a crazy dream, but hey—this is a country of dreamers. Dreamers and Christians.

HELP US, BACKERS, YOU'RE OUR ONLY HOPE!:
Just a little Star Wards humor for ya!

"LOL!":
Live A Little! Give A Lot ("GOL")!

**PLEASE PLEDGE!:**
We've tried pretty much everything else at this point: war, selling some cars, literally making more money (you'd think that would work!!), blaming people, blaming gay "people," war, and debt. None of that has touched the debt. Except debt, which has made the debt worse. Also, war!

If we're able to meet our Kickstarter goal, you will have literally been part of a miracle. A miracle in the great Judeo-Christian tradition of this fair country. Of dreamers!

(let's get on this, people), you will not only receive Special Thanks in the State of the Union as well as on the back of all nickels minted this year at the Denver mint, but Stephanie, our videographer, will take you to a Foster the People concert on October 28, and yes, you can tell people it's a date. This concert's going to be great! They're going to play "Pumped Up Kicks" fourteen times and then, maybe if you clap enough, they'll do an encore and it will be "Pumped Up Kicks"!

**PLEDGE $100,000 OR MORE**
Well HEY there, Mr. Hollywood producer! A pledge of $100,000 or more will get you a walk-on role in the next meeting of Congress. Your vote for bills and propositions will be legally binding, so have fun! Don't name any public parks after racial slurs!

**PLEDGE $1,000,000 OR MORE**
Here's Louisiana.

**PLEDGE $1,000,000,000 OR MORE**
We will change the American flag! We will replace each star with your face. Unless you're black, since the stars are white and it really makes sense for them to be as close to white as possible. Though if you're giving a billion dollars, we'll assume you're white or maybe a Dubaian light brown!

**PLEDGE $1,000,000,050 OR MORE**
Everything from the $1,000,000,000 level, plus a tote.

**PLEDGE $1,000,000,000,000 OR MORE**
Start whatever war you want! You want a new Civil War where Asians have to fight their brown Asian brothers? You got it, bucko! Hate-crime Polacks? A PLUS+++. You want to take over Ireland? Those Polacks haven't done anything in years. JUST want to kill Disney Channel kid stars? THAT'S PRACTICALLY LEGAL ALREADY! Kill a girl. Kill a kid. You got it.

**PLEDGE $20,000,000,000,000 OR MORE**
Kill so many kids. (KOL!)

# Electronic Music

Music is everywhere. It's in our bones. It's in our skin. Just like blood! Which is also everywhere, it seems like! Wait I think my coma-boyfriend is bleeding all over the house again. Brb! **FIG. 7.12**

**FIG. 7.12**

Back! Many famous musicians have gone on record saying they "love music!" *Electronic music* is a relatively new art form that employs electronic musical instruments and technology (such as computer synthesizers) in its production. Electronic music is so cool that it can be hard to keep up with. These are my ten fave/fun and flirty new genres of electronic music!

## 10 NEW GENRES
### *of*
## ELECTRONIC MUSIC

| 1 | 2 | 3 | 4 | 5 |
|---|---|---|---|---|
| Hip house | Clap bass | Jingle core | Skin | Patty ska |

| 6 | 7 | 8 | 9 | 10 |
|---|---|---|---|---|
| Chap | Ennuicore | Unicore | Chints | Gregorian filth |

## Space & Technology Recap

OMG I know we're so close to the end but I just have to digress for like ONE more sec!!!! BEAR WITH ME, BECAUSE . . .

## My Big News Recap

I HAVE THE BIGGEST FORKING NEWS BABY BUBS BUT NO NO I SHOULD JUST FINISH THIS GODDAM CHAP

## Space & Technology Recap

No no no no news time

## My Big News Recap

No no no no space & technology recap time

## Space & Technology Recap

No

# My Big News Recap

OKAY, YOU CONVINCED ME, HERE IT FUCKING IS—MY BOYFRIEND ASKED ME TO MARRY HIM!!!!!!!!!!!!!!!!!!!!!!!!!!!!!!!!!!!!!!!!!!!!!!! IT'S HAPPENING!!!!!!!!!!!!!!!!!!!!!!!!!!!!!!!

Oh my God here's how it happened. It was so romantic, it was so classic. Oh my God I can barely type. I'm about to pass out. Oh my God I just did! Between those sentences I just wrote, I was out cold for like six hours! Anyway okay so I knew something was up when he fell off his gurney onto one knee. Then, as I was helping him back up to the gurney and checking to see that he was still breathing, this "ring" made out of IV tubing fell out of his pocket! It's so cool how he tailored his proposal to how we met (while we were both on IVs). **FIG. 7.13** Okay, I just passed out again for like a full day or two (???). We don't need a ring that's worth one month of his income yet—this is all I need. He's just so sweet, I don't know how I'm going to ever live up to this proposal. There I go again! Another five hours unconscious on the floor!

I've already dived fully into planning my wedding. Between writing this book and passing out so much, I just don't know when I'll have the time. But if this guy who doesn't have a name (note to self: pick cute last name for boyfriend before the wedding so that you can change your name to it!) was able to multitask and juggle being in a *coma* with proposing to me, I'll be able to juggle this book and this wedding.

And the BEST part? YOU'RE ALL INVITED! Even Xander, because I am the bigger person! Figuratively—Xander has really gained some weight!!

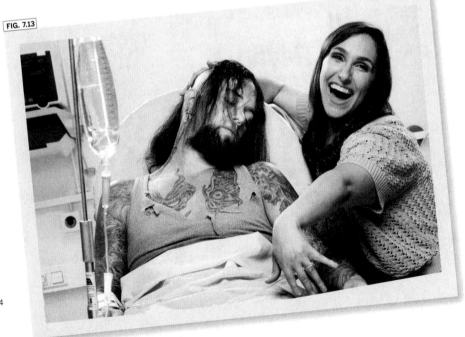

FIG. 7.13

# Recap Questions

**QUESTION 1:** What last name should I give my completely unnamed coma fiancé?

**QUESTION 2:** What size dress do you wear, so that I can send you a bridesmaid dress because you're a bridesmaid!

**QUESTION 3:** Are you sure you really wear that size? You seem like you may have gained a little weight recently, not trying to neg you, just being honest!

**QUESTION 4:** The rotation rate of stars can be determined through spectroscopic measurement, or more exactly determined by tracking the rotation rate of starspots. Young stars can have a rapid rate of rotation greater than 100 km/s at the equator. The B-class star Achernar, for example, has an equatorial rotation velocity of about 225 km/s or greater, giving it an equatorial diameter that is more than 50 percent larger than the distance between the poles. (Wikipedia) What is the critical velocity that would have broken that star apart?

**QUESTION 5:** WILL YOU MARRY ME? JK YOU CAN'T I'M ENGAGED!

# 8

# Women in Science

# Last Chapter, Babes, OMFG

This is the last real chapter. I am going to literally start sobbing and throwing up the tears that I accidentally swallow. I'm going to literally be crylimic. But first: some important business.

Okay, I never thought my news was ever going to be able to ever top the engagement news from last chapter. But honestly, this is probably the craziest thing that has ever happened to me or anyone you or I know. This is so much bigger than the sack dress Uma Thurman wore to the Oscars. FIG. 8.1

Okay, this is really tough to say. I don't know how to start this. But I've realized that I was being held hostage. By my beliefs. And also literally.

FIG. 8.1

I have a confession to make, babes: I have been writing the last couple of paragraphs in a basement. Yes—a basement. I'm pretty sure it's my creepy male neighbor's, but I'm not sure. I've been here for what might be weeks, might be years. One day I'm mowing my lawn post-engagement and the next, I have caked-on blood on my caked-on foundation. In retrospect, I guess I was asking for it. The way I dressed (sweats used for mowing the lawn), FIG. 8.2 there's no way that my creepy neighbor was to know that I *didn't* want to be kidnapped and put into a basement for, oh, I don't know, like seven years?

FIG. 8.2

I want to apologize for not being more up-front about this during the first few sentences of this chapter. It was extremely "bitchy" of me to keep this from you, my *best friends*. I had nothing to be afraid of. You've been extremely supportive so far. But I guess I've hit rock bottom. Figuratively and literally. The bottom of this cage I'm in is made of jagged rocks.

Today is the start of a new Megan. One who tells her best friends (like my best friend Francesca, who has natural blond hair) FIG. 8.3 everything, yes, but also a new Megan who tries to break out of this basement dungeon. If not for me, for my true love, the coman! No one's fed him for months or years—I bet he's pretty pissed off!

FIG. 8.3

However, before I escape, I would like to let all the girls in my position (specifically "Basement Sex Dungeon" = this chapter's sex position!!) know that you don't have to just sit back and take it. You can truly change your environment! By sprucing up the dungeon area that you're being kept in!

# "Trading Dungeons"
## *How to Spruce Up Your Basement Dungeon*

← BEFORE

You can use whatever you find strewn around your surroundings to spice up a boring old dungeon and bring it into the twenty-first century!

AFTER ⟶

← BEFORE

Did your captor leave any cigarette butts and fishing wire around the dungeon? String them together into a wind chime/mobile! This will be especially fun if you give birth to a child in captivity.

AFTER ⟶

← BEFORE

Did they leave chain links to attach you to the wall? Time for a new necklace!

AFTER ⟶

← BEFORE

Use the fingernails that you've scraped off trying to escape as potpourri!

AFTER ⟶

CHECK IT OUT!

# Women in Science: Introduction

"Women in science" comes from the English *women in science*, which, loosely translated, means "bitches in science."

Every *chapter* (segments of the book you've been reading) thus far has somewhat been about women in science, but we're really going to focus on the superstars now. Until the late twentieth century, women were not major parts of scientific findings because they were not usually allowed to be scientists. The few women who were able to become scientists and doctors before the 1900s were probably lesbos. That way they wouldn't distract the male scientists, which was the number-one cause of work-related injuries at that time. Whenever a woman would walk into a laboratory, a man would bend her over a countertop so quickly that they'd spill acid on themselves. Women could not be trusted.

However, a few homely women were able to break through the gender barrier! *Marie Curie*, whom we've already discussed, was a Nobel Prize winner, but her husband probably did all her work for her. They won the Nobel Prize for their work on radiation. Her daughter *Irene* also won a Nobel Prize! Seems like the Curie gals were getting a little Nobel-greedy. Marie Curie actually died of *aplastic anemia* (← this chapter's other sex move), which she got by spending so much time in her radiation lab. Hmm, seems like a little *hint*, hmm? Women shouldn't be showing up their husbands or captors because then they'll get fatal diseases? That's what I tell myself at least!

*Elizabeth Blackwell* was the first doctor. And she wasn't even a vagina doctor! She was just a general ol' doctor! She may have even had to fix a broken dick! FIG. 8.4 I personally would never trust a woman to fix my broken dick. Our hands aren't deft enough for *skin that soft*, or, if you're doing it right, *skin that hard*. I'd never trust a female doctor. Maybe it's just me, but I think women doctors are bad at medicine and I'm caught in a basement dungeon.

FIG. 8.4

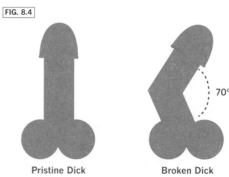

Pristine Dick　　　　Broken Dick

*Jane Goodall* gained notoriety as a *primatologist*, someone who studies primates, such as chimpanzees. She is considered to be the world's foremost expert on chimpanzees. For forty-five years, Goodall studied the social and family interactions of wild chimpanzees in Tanzania. More like she studied *boyfriends, AM I RIGHT, LADIES!!* FIG. 8.5 Hahahahahaha, jk, nothing matters, who fucking cares if your boyfriend is shitty. Deal with it because you are lucky to not be subterranean. You are an idiot if you're wasting time learning about women in science. Go fuck a jock or something.

WHOA! Sheesh, I'm sorry I'm in such a bad mood! I must be way more upset about this whole captive-in-a-basement thing than I thought. I assumed it was PMS but I guess I'm actually upset that I'm caught in a basement??? I have been telling myself I'm fine, but I've got this nagging feeling that it's actually really putting me

out. This little tickling feeling on the roof of my mouth. Oh wait, that's just a spider in my gruel.

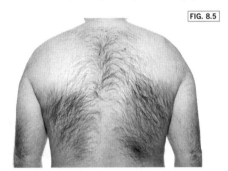

FIG. 8.5

## How to Tell if You're Upset Because You're PMS-ing or Because You're Caught in a Basement Dungeon

**Take this fun, flirty quiz to see whether you're acting "that way" because of your monthly cycle or because you've been trapped underground for any number of months or years, you can't tell time passing anymore!**

 OR

Are you caught in a basement dungeon?
  **A.** Not yet, though inevitably I will be at some point since women are just weak targets to be picked off by male sexual predators!!

  **B.** Yes!

*Mostly A's*
You have PMS!

*Mostly B's*
You're upset about being caught in a basement dungeon!

# Hot or Not?!

Okay, sure, we've gone over some women scientists who've won Nobel Prizes blah blah blah. But what about the women scientists who were *hot*?! That's what I wanna see!

## THE RACHEL

**Go to your hairdresser (mine is my best friend Carlita, she'll dye your hair seriously any color, even brown) and ask for "the Rachel"! She'll know exactly what to give you—the haircut that environmentalist Rachel Carson wore!**

# Famous Women Scientists...
# WITHOUT MAKEUP!

**Even famous female scientists sometimes let their hair down! These ladies are fearless, putting themselves out there without makeup!**

with makeup

MARIE CURIE

without makeup

with makeup

ELIZABETH BLACKWELL

without makeup

with makeup

JANE GOODALL

without makeup

# Men in Women in Science

I know what you're thinking. Any woman can be a scientist, but it takes a *special* person to be a woman scientist who's *married*. Marie Curie actually worked with her husband, *Pierre Curie*. Doctor *Gerty Cori* shared the 1947 Nobel Prize for medicine with her husband, *Carl Ferdinand Cori*, for researching the *conversion of glycogen*. *Mrs. Claus*, worked alongside her husband, *Santa Claus*, in the field of *toys*. Obviously these female scientists weren't hideously ugly if they were able to land successful scientists for husbands! Or maybe they were but their husbands had blinded themselves with chemicals from the lab! That's pretty much the best-case scenario for a female scientist! That the man who works in the laboratory with you maims himself so severely that he can't do anything but marry you, but only if your face feels like you're at least a 6 in attractiveness when he does the thing where he feels your face to "see" what you look like!

# Women with Jobs?!

This chapter brings up a broader topic to discuss—women and employment. I'm sure it's been weighing on your mind throughout the whole book. Should women even *have* a job? It's one of history's greatest unsolved mysteries. The women I just mentioned were able to have successful careers in the sciences, but *should* they have? Call me old-fashioned, but I think women should be housewives only and also I might have smallpox from this basement cave. I truly believe that women should not work outside the home. It's *Steve* Jobs, not *Eve* Jobs.

Women evolved to take care of the young and feed their families. This is irrefutable *fact*. But that doesn't preclude some women from leaving the home (Mrs. Claus, I'm looking at you). They venture out into the world and leave their families behind. It's a tragedy on an epic scale. Shouldn't being your best girlfriend's best friend be job enough for you? My best friend Emma put "best friend" as her profession on her tax returns! In fact, she put "best friend" in every box on her tax returns! She's so adorable. I have to go visit her in prison once I get out of here. FIG. 8.6

I have to say, my opinion on the matter *has* changed a little since I was locked in a basement. That's probably why I didn't write about it before. If I had never ventured out into the yard in my sexy pajamas, I never would have ended up here. That's why I've always admired women who

FIG. 8.6

stay in their homes the most. You might be thinking that I'm a hypocrite, because writing this book could be considered a professional writing job. But, in my defense, I am in a basement now! A woman's place is in the home! Specifically, the basement of someone else's home!

Sure, the women I've discussed in this chapter had jobs that helped the human race and garnered them acclaim. But remember, there's ALWAYS a place in a home for you, especially if your captor is really stepping up and being a man by fully supporting you! Fuck you, Mrs. Claus, your children are going to grow up as hoodlums!!!!!! **FIG. 8.7** I feel so weak those exclamation points exhausted me please give me more water Daddy.

FIG. 8.7

# Women in Science Recap

Sorry this chapter wasn't longer. I'm feeling very weak these days. I'm pretty much only fed rotten bread crusts, which is actually pretty good for my figure. I miss my coman so much that I drew his face on the rock wall using the blood of a rat I found. If I don't make it out, tell my best friend I love her. I mean my *real* best friend. You know which one.

# Recap Questions

**QUESTION 1:** Will you tell my ex-fiancé that I still love him?

**QUESTION 2:** Can you throw me some non-moldy scraps?

**QUESTION 3:** Are you even reading this? If I never leave captivity then it seems doubtful that this manuscript will ever see the light of day. Holy fuck that's depressing. Goddammit.

**QUESTION 4:** I'm so weak, I just can't finish this I'm

Creamed kale!!!! Steamed kale!!!! Grilled kale!!!! Roast kale!!!! Kale chips!!!! Spicy kale chips!!!! Crispy kale chips with sea salt!!!! Spicy kale chips with sea salt!!!! Kale pesto!!!! Kale quiche!!!! Kale soup (hot)!!!! Cold kale soup!!!! Warm kale soup!!!! Kale lasagna!!!! Kale juice!!!! Kale slaw!!!! Kale pasta!!!! Kale pizza!!!! Braised kale!!!! Japanese-Style Boiled Kale Salad!!!! Baked Kale Mac and Cheese!!!! African-Style Stewed Kale!!!! Sautéed Kale with Toast and Eggs!!!! Tropical Kale Slushy!!!! Kale Flatbreads!!!! Sautéed kale!!!! Sautéed kale with olive oil!!!! Sautéed kale with olive oil and garlic!!!! Sautéed kale with olive oil, garlic, and salt!!!! Sautéed kale with olive oil, garlic, salt, and pepper!!!! Sautéed kale with olive oil, salt, and pepper!!!! Sautéed kale with olive oil, garlic, and pepper!!!! Sautéed kale with garlic, salt, and pepper!!!! Sautéed kale with garlic and salt!!!! Sautéed kale with garlic and pepper!!!! Sautéed kale with olive oil and salt!!!! Sautéed kale with olive oil and pepper!!!! Sautéed kale with salt and pepper!!!! Kale-Spiked Mashed Potatoes!!!! Sautéed Tuscan Kale!!!! Smoky Kale Chiffonade!!!! Braised Rice Beans with Dinosaur Kale and Heirloom Tomatoes!!!! Asparagus and Baby Russian Red Kale Slaw!!!! Minestrone soup (with kale)!!!! Potato kale soup!!!! Potato kale stew!!!! Kale cheese pie!!!! Kale smoothie!!!! Raw kale with pine nuts!!!! Grilled Coconut Kale!!!! Creamy Low-Calorie Kale Dip!!!! White Bean Soup with Sausage and Kale!!!! Lemony Kale Salad with Tomatoes!!!! Kale–Goat Cheese Frittata!!!! Beans and Latin Greens with Mojo!!!! Soba Noodles with Kale!!!! Roasted Kale and Red Onions!!!! Kale Mint Smoothie!!!! Thai Kale Slaw!!!! Kale frittata!!!! Kale gratin!!!! Slow-simmer kale!!!! Medium-simmer kale!!!! Fast-simmer kale!!!! Crazy-fast-simmer kale!!!! Indian spice kale!!!! Argentinean spice kale!!!! South Indian spice kale!!!! Mexican spice kale!!!! Algerian spice kale!!!! Hawaiian spice kale!!!! Egyptian spice kale!!!! Brazilian spice kale!!!! Canadian spice kale!!!! Icelandic spice kale!!!! Finnish spice kale!!!! Democratic Republic of the Congo spice kale!!!! Republic of the Congo spice kale!!!! Belgian Congo spice kale!!!! Kale with Citrus Garlic Sauce!!!! Kale with Cashew Sauce!!!! Kale with Creamy Coconut Dressing!!!! Kale with Chunky Coconut Dressing!!!! Kale with Caesar dressing!!!! Kale with ranch dressing!!!! Kale with low-fat ranch dressing!!!! Kale with extra-fat ranch dressing!!!! Bean soup with kale!!!! Sweet & Savory Kale!!!! Winter kale hash!!!! Summer kale hash!!!! Spring kale hash!!!! Autumn kale hash!!!! A Midsummer

Night's kale hash!!!! Sweet Pepper Pasta Toss with Kale!!!! Sour Pepper Pasta Toss withOUT Kale JUST KIDDING!!!! Savory Kale, Cannellini Bean, and Potato Soup!!!! Stir-fried kale!!!! Orzo with kale!!!! Kale-and-banana smoothie!!!! Kale-and-strawberry smoothie!!!! Kale-and-apple smoothie!!!! Kale-and-pear smoothie!!!! Kale-and-pineapple smoothie!!!! Kale-and-coconut smoothie!!!! Kale-and-blueberry smoothie!!!! Kale-and-blackberry smoothie!!!! Kale-and-acai smoothie!!!! Kale-and-mango smoothie!!!! Kale-and-cherry smoothie!!!! Kale-and-elderberry smoothie!!!! Kale-and-red-grape smoothie!!!! Kale-and-green-grape smoothie!!!! Kale-and-hazelnut smoothie!!!! Kale-and-persimmon smoothie!!!! Kale-and-plum smoothie!!!! Kale-and-peach smoothie!!!! Kale-and-apricot smoothie!!!! Kale-and-raspberry smoothie!!!! Kale-and-crabapple smoothie!!!! Kale-and-papaya smoothie!!!! Kale-and-avocado smoothie!!!! Kale-and-mulberry smoothie!!!! Kale-and-boysenberry smoothie!!!! Kale-and-guava smoothie!!!! Kale-and-tomato smoothie!!!! Kale-and-gooseberry smoothie!!!! Kale-and-cantaloupe smoothie!!!! Kale-and-clementine smoothie!!!! Kale-and-fig smoothie!!!! Kale-and-cranberry smoothie!!!! Kale-and-currant smoothie!!!! Kale-and-passion-fruit smoothie!!!!

Kale-and-date smoothie!!!! Kale-and-dragonfruit smoothie!!!! Kale-and-lime smoothie!!!! Kale-and-lemon smoothie!!!! Kale-and-orange smoothie!!!! Kale-and-tangerine smoothie!!!! Kale-and-galia-melon smoothie!!!! Kale-and-jackfruit smoothie!!!! Kale-and-lingonberry smoothie!!!! Kale-and-lychee smoothie!!!! Kale-and-macadamia smoothie!!!! Kale-and-muskmelon smoothie!!!! Kale-and-nut-meg smoothie!!!! Kale-and-pistachio smoothie!!!! Kale-and-pomegranate smoothie!!!! Kale-and-pom-elo smoothie!!!! Kale-and-pumpkin smoothie!!!! Kale-and-quince smoothie!!!! Kale-and-rhubarb smoothie!!!! Kale-and-rose-hip smoothie!!!! Kale-and-tamarind smoothie!!!! Kale-and-thimbleberry smoothie!!!! Kale-and-almond smoothie!!!! Kale-and-vanilla smoothie!!!! Kale-and-watermelon smoothie!!!! Kale-and-huckleberry smoothie!!!! Kale-and-kumquat smoothie!!!! Kale-and-loquat smoothie!!!! Kale-and-medlar smoothie!!!! Kale-and-rowan smoothie!!!! Kale-and-serviceberry smoothie!!!! Kale-and-shipova smoothie!!!! Kale-and-chokecherry smoothie!!!! Kale-and-greengage smoothie!!!! Kale-and-apriplum smoothie!!!! Kale-and-goumi smoothie!!!! Kale-and-Lardizabala smoothie!!!! Kale-and-olallieberry smoothie!!!! Kale-and-tayberry smoothie!!!! Kale-and-wineberry smoothie!!!! **Kale-and-bearberry smoothie!!!!** Kale-and-crowberry smoothie!!!! **Kale-and-honeysuckle smoothie!!!!** Kale-and-mayapple smoothie!!!! **Kale-and-nannyberry smoothie!!!!** Kale-and-sea-buckthorn smoothie!!!! **Kale-and-ugni smoothie!!!!** Kale-and-jujube smoothie!!!! **Kale-and-blood-orange smoothie!!!!** Kale-and-citron smoothie!!!! **Kale-and-naartjie smoothie!!!!** Kale-and-kabosu smoothie!!!! **Kale-and-oroblanco smoothie!!!!** Kale-and-carob smoothie!!!! **Kale-and-feijoa smoothie!!!!** Kale-and-longan smoothie!!!! **Kale-and-Lúcuma smoothie!!!!** Kale-and-peanut smoothie!!!! **Kale-and-pond-apple smoothie!!!!** Kale-and-tamarillo smoothie!!!! **Kale-and-yangmei smoothie!!!!** Kale-and-Néré smoothie!!!! **Kale-and-acerola smoothie!!!!** Kale-and-ackee smoothie!!!! **Kale-and-African moringa smoothie!!!!** Kale-and-agave smoothie!!!! **Kale-and-allspice smoothie!!!!** Roasted yam and kale salad!!!! **Kale, quinoa, and avocado salad with lemon Dijon vinaigrette!!!!** Kale, quinoa, and avocado salad with ranch dressing!!!! **Kale, quinoa, and avocado salad with Caesar dressing!!!!** Kale, quinoa, and avocado salad with lite Caesar dressing!!!! **Kale, quinoa, and avocado salad with Thousand Island dressing!!!!** Kale, quinoa, and avocado salad with oil and vinegar!!!! **Kale, quinoa, and avocado salad with bleu** cheese dressing!!!! Kale, quinoa, and avocado salad with extra virgin olive oil!!!! Kale, quinoa, and avocado salad with Italian dressing!!!! Kale, quinoa, and avocado salad with French dressing!!!! Kale, quinoa, and avocado salad with ginger dressing!!!! Kale, quinoa, and avocado salad with hummus!!!! Kale, quinoa, and avocado salad with Louis dressing!!!! Kale, quinoa, and avocado salad with Russian dressing!!!! Kale, quinoa, and avocado salad with tahini dressing!!!! Kale, quinoa, and avocado salad with tahini dill dressing!!!! Kale, quinoa, and avocado salad with wafu dressing!!!! Kale, quinoa, and avocado salad with soy sauce!!!! Kale, quinoa, and avocado salad with low-sodium soy sauce!!!! Kale, quinoa, and avocado salad with sodium-free soy sauce!!!! Kale and quinoa salad!!!! Kale krisps!!!! Kale Puttanesca!!!! Stir-fried kale and broccoli florets!!!! Bean-and-Kale Ragu!!!! Kielbasa Kale Stew!!!! Portuguese Kale Soup!!!! Kale and Adzuki Beans!!!! Sister Slaw with Kale!!!! Kale with Cannellini Beans and Pancetta!!!! Lisa's Co-op Kale Soup!!!! Chili-Roasted Kale!!!! Kale, Swiss Chard, Chicken, and Feta Salad!!!! Fast and Easy Kale Soup!!!! Pork Tenderloin with Steamed Kale!!!! Vegan Lentil, Kale, and Red Onion Paste!!!! Pan-Fried Polenta with Corn, Kale, and Goat Cheese!!!!

Super Summer Kale Salad!!!! Indian Summer Kale Salad!!!! Groundhog Day Kale Salad!!!! West African Peanut Stew with Kale!!!! East African Peanut Stew with Kale!!!! North African Peanut Stew with Kale!!!! South African Peanut Stew with Kale!!!! Northwest African Peanut Stew with Kale!!!! Southwest African Peanut Stew with Kale!!!! Southeast African Peanut Stew with Kale!!!! Northeast African Peanut Stew with Kale!!!! Moroccan Chickpea Stew with Kale!!!! Kale and Quinoa with Creole Seasoning!!!! Black-Eyed Peas with Pork and Greens (kale)!!!! Dawn's Kale Side Dish!!!! Cannellini Beans with Flat-Leaf Kale!!!! Sautéed Rice with Kale!!!! Lemony Lentils with Kale!!!! Ohio Sausage and Kale Soup!!!! Mickey L's Kale Soup!!!! Creamy Kale Salad!!!! Kale, Seitan, and Mac-and-Cheese Sandwich!!!! Tofu-Eggless Salad with Kale!!!! Braised Kale with Bacon and Cider!!!! Garbanzo Beans and Kale!!!! Spinach-and-Kale Turnovers!!!! Farfalle with Sausage and Kale!!!! White-Bean-and-Sausage Ragout with Tomatoes, Kale, and Zucchini!!!! Bacon and Butternut Pasta with Kale!!!! Braised Chicken with Kale!!!! Poached Chicken with Kale!!!! Fried Chicken with Kale!!!! Pasta with Black Kale, Caramelized Onions, and Parsnips!!!! White Bean Soup with Kale and Chorizo!!!! Sausage and Clams with

Chickpeas and Kale!!!! Orecchiette with Kale, Bacon, and Sun-Dried Tomatoes!!!! Two-Bean Soup with Kale!!!! Borlotti Minestrone with Kale!!!! Cajun Steak Frites with Kale!!!! Wilted Kale with Bacon and Vinegar!!!! Wilted Kale with Fake Vegetarian Bacon and Vinegar!!!! Wilted Kale with Coconut, Ginger, and Lime!!!! Wilted Kale with Farro and Walnuts!!!! Wilted Kale with Golden Shallots!!!! Wheat Berry, Kale, and Cranberry Salad!!!! Turkey Meatball Soup with Wilted Kale!!!! Kale-and-Caramelized-Onion Grilled Cheese!!!! Cauliflower and Kale with Mustard Currant Dressing!!!! Paleo Chorizo Kale and Sweet Potato Soup!!!! Kale and Spinach Saag!!!! Chicken and Kale in Parmesan Cream Sauce!!!! Rustic Tuscan Soup with Kale!!!! "Amazing" Lentils and Kale!!!! Kale and Feta Salad!!!! Curry Kale and Potato Galette!!!! Italian Ribollita (Vegetable and Bread Soup) with Kale!!!! Slow Cooker Chicken Chili with Kale and Beans!!!! Pesto Spaghetti Squash and Kale!!!! Pumpkin, Kale, and Black Bean Stew!!!! Kale Couscous!!!! Kale Salad with Pomegranate, Sunflower Seeds, and Sliced Almonds!!!! Kale Chips with Honey!!!!!!!!!!!!!!!!! !!!!!!!!!!!!!!!!!!!!!!!!!!!!!!!!!!!!!!!!!!!!!!!!!!!!!!!!!!!!! !!!!!!!!!!!!!!!!!!!!!!!!!!!!!!!!!!!!!!!!!!!!!!!!!!!!!!!!!!!!! !!!!!!!!!!!!!!!!!!!!!!!!!!!!!!!!!!!!!!!!!!!!!!!!!!!!!!!!!!!!!

# Introduction

Here it is. The end. I knew it was coming, but it's still a bit of a shock. It's like Xander—I guess a little part of me always knew the end was inevitable after he told me we were breaking up. I just didn't know when. And for our relationship, it happened to end right after he said, "This is over, Megan." I had to learn that that's guy code for "this is over."

And it's like my unnamed fiancé. The end came for us when I was kidnapped from my front yard by a strange man who was probably my neighbor. It was a very bittersweet ending to our love. The "bitter" part was, you guessed it, being kidnapped.

I've obviously been through a lot. The Xander stuff. The unnamed coman fiancé stuff. The dungeon stuff. But something happened to me recently that was SO good that I feel like I'll never have to have anything good ever happen to me again. It's so good that I'm not even crying, I'm just sitting and steeping in my own happy-juices (last chapter's featured sex move! OMG I can't believe I'm saying "last chapter"!).

Okay, so here's the news. I was not able to break out of the dungeon. I hate to say "never," so I'll try to break out again after my yogalates has progressed and I've worked up my triceps and delts a bit. BUT everything is LOOKING UP for ol' Megan—my luck is truly incredible. I have to thank G-d because someone up there is smiling on me. Get this—my captor turned out to be a PUBLISHER! I CAN'T BELIEVE MY LUCK! AHHHHHHHHHHHHHHHHHHH-HHHHHHHHHHHHHHHHHHHHHHHHHH-HHH!!!!!!!!!!!!!!!!!!!!!! What are the CHANCES that the ONE guy who would keep me in his basement for seven years would ALSO HAPPEN to be a successful publisher in the print-textbook industry????? I am so grateful I just can't believe it.

I gave him my manuscript and after some editing and back-and-forth and heavy beatings we were able to agree on a final draft that he felt proud to secretly publish under a pseudonym. I was so flattered, I may not try to escape again. I think maybe we should just get married!! Hey, he captured me, he deserves it! It's a "you break it, you bought it" situation!

But this conclusion isn't about my dungeon-boyfriend. It's about *us*. Girl, we have been through so much together as best friends. We have changed *so* much. I don't even recognize the girl who wrote "You are my best friend." But crazily enough, that was me, in the first chapter of this book!

We have had our ups and downs. I for one have grown *so* much as a person. At the start of this science journey, I was a lonely NASA employee waiting for Xander to come crawling back. I would have never imagined that, at the end of this whole thing, I'd ever be in a publisher's dungeon, let alone *engaged* to him! Eesh, I shouldn't talk like that. We haven't even discussed marriage, that was just an idea I literally just had. I am so clingy! How long do you wait to text the guy who's keeping you in his dungeon about marriage so he doesn't think you're clingy?

And look at YOU! You're barely recognizable! When you entered into this tacit pact with my book, you were ignorant, shallow, a fucking bitch. Now you're basically a genius. You know

everything. You got highlights. You have scallops for lunch (no carbs!). You can build an X-ray. You can capture the sun. You are a real woman.

No matter where your life takes you, remember that science can only make it richer. Whether you eventually get married or you eventually get married to the person whose dungeon you're in, science will accompany you forever.

*In conclusion, I will love you forever. And again, here is an address where you can send me, Megan Amram, checks:*

Megan Amram c/o Scribner Publishing, 1230 Avenue of the Americas, New York, NY 10020.

## Not Pictured:

Carly, Michelle, Ali, Nicki, Ashley, Katie, Candy, Ellen, Miranda, Eliana, Chloe, Olive, Mel, Amanda, Davida, Mandy, Alexis, Carrie, Rachael Ray, Tiffany, Heidi, Minnie, Mary Katherine, Kristina, Nora, Reverse Cowgirl, Raita, Marie Claire, Clara, Adele, Aisha, Katja, Lauren, Claire-Marie, Kathleen, Crystal, Crystal Glass, Christina, Tina, Cris, Cristy, Ice, Getgo, G, Trash, Super Ice, LA Glass, LA Ice, Ice Cream, Quartz, Chunky Love, Cookies, No Doze, Pookie, Rocket Fuel, Scooby Snax, Rebecca, Marissa, Carla, Ella, Aubrey, Lily, Bridget, Sofia, Hannah, Amelia, Arianna, Harper, Lillian, Charlotte T., Charlotte A., Charlotte M., Evelyn, Victoria, Brooklyn, Zoe, Layla, Hailey, Leah, Kaylee, Riri, Gabriella, Alison, Shirlene, Nancy, Carlita, Rachel Carson, Maddie, Lizzy, Sophie, Robyn, Natalie, Alexandra, Francesca, Maggie, Pilar, Claudia, Sasha, Rachna, Rakhee, Jen, Alyssa, Sophia, Adina, Ava, Salom, Hamm, Isabel, Abigail, Mia, Madison, Elizabeth, Avery, Addison, Mackenzie, Giana, Faith, Melanie, Blanche, Sydney, Bailey, Caroline, Naomi, Morgan, Kennedy, Lindsay, Audrey, Savannah, Alissa, Claire, Taylor, Riley, Camila, Brianna, Rheeqrheeq, Peyton, Bella, Meg, Alexa, Kylie, Kira, Dereka, Benita, Max, Scarlett, Stella, Maya, Catherine, Julia, Lucy, Madelyn, Autumn, Summer, Ellie, Jasmine, Chris, Skylar, Kimberly, Violet, Molly, Aria, Jocelyn, Trinity, London, Lydia, Annabel, Jessica, Jennifer, Jaycee, Stephanie Sondheim, Angie, Brittany, Nicole, Heather, Barrett, Samantha, Melissa, Danielle, Amber, Maxine, Laureen, Kim, Laura, Amy, Kayla, Katherine, Sara, Kelly, Erica, Whitney, Courtney, Erin, Angela, Jan, Andrea, Jamie, Lisa, Tammy, J.J., Julie, Dawn, Karen, Susan, Christine, Cynthia, Lori, Patricia, Pamela, Wendy, Sandy, Stacy, Debbie, Nita, Carolyn, Bernice, Betsy, Janice, Shannon, Kit, April, Lesley, BenSimone, Lindsey, Kristin, Roberta, Ezri, Blostam, Edna, Bradena, Alicia, Donna, Rose, Petra, Aparna, Augusta, Audra, Ronnie, Jonna, Artis, Natasa, Billie, Ashton, Mary Ellen, Tricia, Kara, Mary-Todd, Bridga, Kiki, Sammi, Aleks, Juliet, Maria, Lolo, Mrs. A, Celine, Mrs. L, Kathryn, Tara, Magda, Monica, Jacqueline, Holly, Cassandra, Brandy, Chelsea, Brandie, Leslie, Diana, Dana, Jenna, Brooke, Matilda, Valerie, Caitlin, Stacey, Brittney, Margaret, Sandra, Tali, Joanne, Phyllis, Lucille, Candice, Nasia, Meghan, LaToya, Bethany, Misty, Katrina, Karey, Kelsey, Joy, Jillian, Denise, Sabrina, Gina, Jill, Eryn, L.W., Gregoria, Daniella, Alana, Michaela, Bennie, Marina, Donica, Sam, Harissa, Naftalia, Golda, Norma, Dani, Phillipa, H.P.T., Jackie, Jane, Fluffy, Branty, Alex, Carol

# FINAL EXAM:

*Test your comprehension of* Science . . . *for Her! with this fun and flirty final exam!*

## Dedications

**1. Which of these is *not* one of my best friends?**
- **A.** Donica
- **B.** Alana
- **C.** Katie
- **D.** None of the above, every girl is my best friend!!!!!! If you don't choose this answer you're a big dumb idiot JK I LOVE YOU!!

## Biology

**2. What do I smell like when I look at great jeans?**
- **A.** Gas station burrito filled with cottage cheese
- **B.** Dead caterpillar filled with saag paneer
- **C.** 7-Eleven meatball sub filled with rotten Starburst
- **D.** Ikea meatball filled with squid ink and horse hair

## Chemistry

**3. What was Marie Curie in the right light?**
- **A.** A "7"
- **B.** A "4"
- **C.** A "9"
- **D.** A "10" (yeah, right, babe, keep dreaming ;))

## Physics

**4. The word *physics* is roughly translated from which Greek word?**
- **A.** πρύμνη
- **B.** φυσική
- **C.** βαρέλι
- **D.** ταχύτητα

## Chapter Five

**5. What page was cited from *Twilight Saga: Breaking Dawn*?**
- **A.** 132
- **B.** 133
- **C.** 134
- **D.** 135

## Pharmacology & Medicine

**6. Can you get pregnant from legitimate rape?**
- **A.** Of course not. Don't be a big dumb idiot!!!!!
- **B.** Yes, rape often results in pregnancy and can be an extreme physical and psychological trauma.
- **C.** Yes, rape often results in pregnancy and can be an extreme physical trauma but not a psychological one.
- **D.** There's no such thing as rape. It's an urban legend, you scaredy-cat!

## Space & Technology

**7. How do you pronounce "$$$"?**
- **A.** Shhhhhhhuhhh
- **B.** Samantha
- **C.** Sssssstchkzh
- **D.** Hannukah

## Women in Science

**8. What did doctors Gerty Cori and her husband Carl Ferdinand Cori share the 1947 Nobel Prize in medicine for?**
- **A.** Work on serum therapy, especially its application against diphtheria
- **B.** Researching the conversion of glycogen
- **C.** Stupidest Name Award Nobel Prize
- **D.** For the discovery of how chromosomes are protected by telomeres and the enzyme telomerase and Stupidest Name Award (first double-award Nobel Prize winners)

## KALE!!!!

**9. Kale?**
- **A.** Kale!!
- **B.** Kale!!!!!
- **C.** Kale!!!!
- **D.** Kale!!!

## Conclusion

**10. How much money should you send to me in check form at "Megan Amram c/o Scribner Publishing, 1230 Avenue of the Americas, New York, NY 10020"?**
- **A.** None, you don't need it
- **B.** Hmm, I dunno, like ten bucks?
- **C.** Hmm, I dunno, like a thousand bucks?
- **D.** Literally everything I can spare, you are such a treasure, Megan, and I cannot thank you enough monetarily

## Final Exam

**11. Which of these is *not* a question on the final exam?**
- **A.** 6: Can you get pregnant from legitimate rape?
- **B.** 4: The word *physics* is roughly translated from which Greek word?
- **C.** 11: Which of these is not a question on the final exam?
- **D.** 476: What's a cool smell for a butt?

**Answers:**

1-d; 2-c; 3-a; 4-b; 5-a; 6-a; 7-c; 8-b; 9-c; 10-d; 11-d

# Index

*I can't barely bear to say this fun and flirty good-bye! You close the book first!! No, YOU close the book first!!!!!*

♥